THE
LAMP AND
LIGHTING
BOOK

Other Books by the Authors:

THE FRAME BOOK
PAPER AS ART AND CRAFT

By Thelma R. Newman:

CONTEMPORARY AFRICAN ARTS AND CRAFTS
CONTEMPORARY DECOUPAGE
CRAFTING WITH PLASTICS
CREATIVE CANDLEMAKING
LEATHER AS ART AND CRAFT
PLASTICS AS AN ART FORM
PLASTICS AS DESIGN FORM
PLASTICS AS SCULPTURE
QUILTING, PATCHWORK, APPLIQUÉ, AND TRAPUNTO
WAX AS ART FORM

By Jay Hartley Newman and Lee Scott Newman:

KITE CRAFT
PLASTICS FOR THE CRAFTSMAN
WIRE ART

By Thelma R. Newman and Virginia Merrill:

THE COMPLETE BOOK OF MAKING MINIATURES

THE LAMP AND LIGHTING BOOK

DESIGNS, ELEMENTS, MATERIALS,
SHADES—FOR STANDING LAMPS,
CEILING AND WALL FIXTURES

Thelma R. Newman
Jay Hartley Newman
Lee Scott Newman

CROWN PUBLISHERS, INC., NEW YORK

TO   *Allie H. Siegel, father and grandfather*

All photographs by the authors unless otherwise noted.

© *1976 by Thelma R. Newman, Jay Hartley Newman, and Lee Scott Newman.*

All rights reserved. No part of this book may be reproduced or utilized in any form or by any means, electronic or mechanical, including photocopying, recording, or by any information storage and retrieval system, without permission in writing from the publisher.

Inquiries should be addressed to Crown Publishers, Inc., One Park Avenue, New York, N.Y. 10016

Library of Congress Cataloging in Publication Data

Newman, Thelma R
   The lamp and lighting book.

  Bibliography: p.
  Includes index.
   1. Electric lamps—Amateurs' manuals. 2. Lampshades—Amateurs' manuels. I. Newman, Jay Hartley, joint author. II. Newman, Lee Scott, joint author. III. Title.
TK9921.N48   621.32'2   75-43564
ISBN 0-517-51862-7
ISBN 0-517-51863-5 pbk.

Printed in the United States of America
Published simultaneously in Canada by
General Publishing Company Limited

Designed by Ruth Smerechniak

Acknowledgments

Writing this book was fun because designing and producing handsome, useful lamps gives enormous satisfaction. Many people aided us along the way by contributing their knowledge and making our research tasks much easier. Among them are Evan Williams of Williams Lamps, in Westfield, N.J., and Pat and Andy Griglak of Lampcrafters, in North Plainfield, N.J. To them goes our appreciation.

Our thanks go, as well, to master glass artist Jack Cushen and designer Curtis Stephens for taking the time and trouble to demonstrate their techniques and share their secrets with others.

The book would hardly have been complete without photographs of fine lamp designs manufactured by many companies. Special thanks go to Stan Angelo, Jr., of LAMPARTS, Angelo Brothers Company. We thank them for aiding us—and for encouraging good design too.

As ever, we are grateful to Norm Smith for fine photo processing, and our thanks go to Pat Weidner for pounding away at the typewriter.

The person who deserves the greatest accolades, however, is Jack Newman, husband and father, who pitched in at every request and continually expedited all processes as needed.

Contents

ACKNOWLEDGMENTS	v
PREFACE	x
1 LIGHT AND DESIGN—PAST AND PRESENT	1
DEVELOPMENT OF LIGHTING	1
PHYSICAL ATTRIBUTES OF LIGHT	12
DESIGNING WITH LIGHT	12
Illumination Levels and Visual Comfort	13
Basic Modes of Lighting	15
Appearance	18
Practicability and Simplicity of Maintenance	18
Suitability of Materials	21
Economy	21
Mood-Creating Aspects	22
2 CONSTRUCTING AND WIRING THE BASIC LAMP	25
LAMP PARTS	26
PARTS COMMON TO MOST TYPES OF LAMPS	27
TABLE LAMP PARTS	31
CEILING LAMP PARTS	34
ELECTRICAL PARTS FOR LAMPS	36
Bulb Sockets	36

	Switches	39
	Plugs	41
	Wire	41
ELECTRICAL WIRING TECHNIQUES		42
SAFETY		43
PREPARING THE WIRE FOR CONNECTIONS		44
MAKING CONNECTIONS		45
	Wire to Screw Terminal Connections	45
	The Solderless Connection	46
	Splicing and Soldering	46
WIRING LAMP PARTS		49
	The Standard Plug	49
	Clip-on Plugs	50
	The Bulb Socket	50
	Series of Sockets	52
	In-Line Switches	53
	Other Types of Switches	54
	Ceiling and Wall Hookups	56
	Fluorescent Fixtures	58

3 LAMPSHADE MAKING — 60
   KINDS OF LAMPSHADES — 60
   DESIGN ASPECTS — 62
   Proportion — 62
   Suitability of Coverings — 63
   Scale — 64
   Function — 64
   Style — 65
   Innovation — 65
   GENERAL ASPECTS OF MAKING A BASIC LAMPSHADE — 65
   Material Selection — 66
   Patternmaking for Drums, Cylinders, and Cones — 67
   Patternmaking for Square, Rectangular, and Hexagonal Shades — 68
   Preparation of the Frame — 69
   Applying Hard Materials — 70
   Applying Soft Coverings — 76
     Pinning — 76
     Stitching — 76
   Tailored Soft-Covered Shades — 79
   Linings — 80
   Trimmings — 80
     Placement of Trims — 83
     Attaching Trims — 83

	VARIATIONS IN LAMPSHADES	84
	Overlays	84
	Decoupage	84
	The Pierced and Cut Look	86
	Unusual Materials	87
	Leather	87
	Copper	91
	Pleated Shades	91
	Forming Pleats	92
	Attaching Pleated Shades	93
	Pleating Variations	94
	Fluted Shades	100
	Slipcover Shades	101
	Wrapped Shades	101
	ABOUT HARPS	103
4	THE PORTABLES: TABLE AND FLOOR LAMPS	104
	WORKING WITH WOOD	105
	Tools	105
	Basic Wood Processes	105
	Gluing Wood	105
	Veneering Wood	106
	Texturing Wood	106
	Finishing Wood	122
	Sanding	122
	Staining	122
	Clear Finishes	123
	PLASTIC TECHNIQUES	123
	Marking and Sawing Acrylic	124
	Drilling and Tapping	125
	Finishing an Acrylic Edge	125
	Scraping and Sanding	125
	Polishing and Buffing	127
	Cementing Acrylic	127
	Heating and Forming Acrylic	141
	MIXED MEDIA	152
5	CEILING AND WALL FIXTURES	156
6	LAMPS FROM FOUND OBJECTS	183
	TECHNIQUES FOR RE-CREATING OBJECTS	184
	Drilling a Hole in Glass or China	184
	Drilling a Hole through Woven Reed and Other Fibers	186
	Forming a Hole through a Gourd	191

	Connecting Attachments	191
	SOME FOND IDEAS FOR FOUND OBJECT LAMPS	193
7	**STAINED-GLASS-TYPE LAMPS**	196
	BACKGROUND	196
	MAKING THE PATTERN	198
	TYPES OF GLASS	200
	CUTTING GLASS	200
	FUSIBLE PLASTICS	208
	CONSTRUCTION AND ATTACHMENT TECHNIQUES	208
	ADHESIVE COPPER TAPE	209
8	**THE LAMP AS SCULPTURE**	216
	LIGHT BOXES	221
	MIRROR LAMP	229
	PIPE LAMP WITH SMOKY BULBS	231
	SUPPLY SOURCES	236
	INDEX	244

Preface

Why make lamps? When you think about it, you can't afford not to. To begin with, commercial lamp prices are often nothing short of astronomical. But that's not even half the reason. Lamps are essential. We need them for light, for decoration, to create an ambience. Lamps provide an inexpensive way to give a room a face lift, and lamps can be constructed to accent and complement any decor.

Even beyond those practical considerations, lampmaking offers an enjoyable leisure activity. The technical aspects are surmounted readily; lamps are easy to make. At one time, lampmaking at home was difficult because it was impossible to find the proper parts. Today, lamp parts are available at hardware stores, lamp stores, and by mail order. And, since the technical elements are so soon mastered, there remains enormous room for experimental innovation and originality. No longer need anyone agonize over not being able to find the proper lamp for a certain spot—make it!

Constructing lamps is what this book demonstrates. From wiring to building bases, using plastic, metal, wood, glass, paper, marble, and found objects, to making lampshades, designing ceiling fixtures, table, wall, and floor lamps, as well as pole lamps. And, while all the technical aspects are dealt with, we hope that no one will just set about copying the designs ad infinitum. Build a few to see what it feels like, and then set out on your own. Experiment with found objects and new materials. It's an endlessly creative and enjoyable craft.

1

Light and Design— Past and Present

DEVELOPMENT OF LIGHTING

Light rules over life: without it people cannot see. Recognizing this, mankind has tried to understand and harness light for millions of years. During most of man's existence, fire provided the only source of illumination—other than the sun. For millennia, fire was the only controllable quantity.

The first attempt to harness fire as a light source was probably an accident when a branch of wood from a fire was used to light the way. Gradually, people learned that certain resinous woods provided the longest-burning and brightest lights and that straight, long splinters were best. Those primitive torches are still being used in isolated parts of the world.

What is more, at least half the world's population still has only flame sources to provide light; while the rest of the world has based its life-style on many man-made lighting forms—incandescent lamps, fluorescent lamps, gaseous vapor lamps, and arc lamps.

Since the torch was the most significant light source for thousands of years, it became much more refined. By the Middle Ages, the sophisticated torch consisted of natural fibers coated with a flammable substance. During that time, pedestrians carried the "flambeau"—as torches were known in France—to light their way at night. But other lighting choices, such as candles and oil lamps, were available as well.

The earliest record of an oil lamp dates back to the Stone Age, when hollow stone lamps were discovered. That container sufficed for thousands of years, until terra cotta was formed into lamps. Terra cotta forms found on the Mesopotamian Plains are thought to be eight to nine thousand years old. The Egyptians and Persians employed copper and bronze lamps circa 2700 B.C., and by 1000 B.C. a vegetable-fiber wick that burned olive oil or nut oil in a saucerlike vessel was used. That was a big advance. It took millions of years for people to reach that stage of sophistication.

By 500-400 B.C., oil lamps were commonly used. But the next step was not achieved until 1784 when a Swiss physicist, Aimé Argand, patented a lamp utilizing a round burner and a tubular wick. He also introduced a chimney for directing and regulating the flow of air to the flame. In 1800, Bertrand G. Carcel improved Argand's oil lamp by adding a clockwork pump to raise oil to the wick. The lamp was so precise that it became a photometric standard for Carcel's day.

Different types of oil, different devices for feeding oil to the flame, and different surface configurations emerged. But, although oil lamps became more attractive, they didn't change much for a century. The kerosene lamp—with its flat wick, perforated metal holder, and plain glass chimeny—was a 19th-century innovation. It remains a familiar source of emergency lighting.

Another source of emergency light—the candle—has an ancient history as well. Although candles did not develop as early as oil lamps, they did become an important light source. Clay candleholders dating from the fourth century B.C. were found in Egypt, and Tut-ankh-amen's tomb (1349 B.C.) contained a candleholder consisting of a bronze socket on a wooden block. It was also reported that Teta, first king of the Sixth dynasty, used candleholders of conventional design. Sir Arthur Evans, reporting his finding of candleholders in the Palace of Minos on Crete, said that the object ". . . was surely intended for a stick of superfine material such as wax." In A.D. 11, Pliny, a Roman historian, discussed a true candle of flax rope soaked in pitch and wax.

For some mysterious reason, there is a gap in candle-using history, from Roman times until it reappeared in prominent use during the Middle Ages. At the beginning of the Reformation (sixteenth century), one church in Wittenberg, Germany, is said to have consumed over twelve tons of candles per year in its 174 candelabra, holding 8,730 candles. Church tithes were payable in beeswax. Until the discovery of petroleum, the candle remained essentially the same: it was made of tallow or beeswax—and it served as the main method of artificial illumination available to the average person.

Looking back, it probably was the torch—soaked in resins and pitch and waxes—that provided the conceptual basis for Western world can-

Oil was one of the earliest materials used for lighting. This is a twelfth/thirteenth-century Syrian oil lamp of green glazed pottery. *Courtesy: The Metropolitan Museum of Art, Bequest of William Milne Grinnell, 1920*

Another early lamp in Damascus ware, a blue and black on white glazed pottery. These lamps were used in Turkish mosques during the sixteenth/seventeenth century. *Courtesy: The Metropolitan Museum of Art, The Theodore M. Davis Collection, 1915*

A brass oil lamp from the Java-Bali area of Indonesia. Made by the lost wax process, this lamp dates back many centuries.

A candleholder from Carfi, sub-Minoan period, eleventh century B.C. *Courtesy: Heraklion Archaeological Museum, Crete, Greece*

dles. There is a distinction, however, between candles and torches, splinters, or rushlights. The latter do not have wicks; the candle does, and it ranks very high in the scale of invention.

Early in Eastern civilizations, the Chinese and Japanese molded candles in paper tubes using the wax of an insect called *cocus* and the seeds of certain trees. Later, the Japanese made wicks of rolled-up rice paper. Their candle could then fit into the sharp spur of an iron or bronze candleholder. In India, candles of animal fat were prohibited by religious decree, but wax was skimmed from boiling cinnamon and made into tapers for temple and ceremonial use.

As candles gained in popularity, their containers became more imaginative. By the fifteenth century, crystal chandeliers were fitted with small cups to hold wax candles. Candelabra of different designs, fitted with candles, lighted homes and public gathering places. One hundred and twenty of these candles generated the light of one modern 100-watt bulb. With the increased popularity of gas lamps during the early nineteenth century, candles were gradually relegated to church ceremonies and festive occasions.

It was not until 1664, near Wigan, Lancashire, England, that John Clayton discovered a pool of natural gas near a coal mine. Still later, gas was extracted from coal by a distillation process. By 1784, Jean Pierre Minckelers learned to harness gas for lighting. But it was not for fourteen more years that gas was used seriously for illumination. In 1798, William Murdock installed gas lighting in an English factory and provided gas commercially to shops, piping it through tubes from its source. Improvements such as the Bunsen burner and the Welsbach mantle emerged in later years. Bunsen created a device—a simple tube of metal with holes at the bottom and top—which emitted a very hot flame. Welsbach developed a cylinder of woven cotton soaked in chemi-

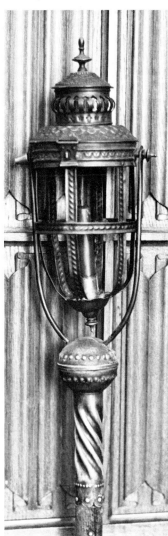

This portable brass candleholder on a velvet-covered shaft dates back to the sixteenth century. This example is at Wroxton College in Wroxton, England (formerly Wroxton Abbey).

Detail of the candle housing shows that the candle always pivots to an upright position even when the shaft is held at an angle. Gimbal fittings such as this are found today on lampshades in England. It is another way of tilting the light source to direct light.

A very old candle lamp with more recent decoupaged background and hand-painted details. By the late Lydia Irwin.

An American clear flint glass lamp, dating back to the early nineteenth century. *Courtesy: The Metropolitan Museum of Art, Rogert Fund, 1918*

A nineteenth-century kerosene lamp of amberina glass from Wheeling, West Virginia. Made by Hobbs, Brocunier & Co. The base is 4 5/8" high, the shade 4 1/2" in diameter and 3" high, the chimney 3 1/2" high. *Courtesy: The Metropolitan Museum of Art, Gift of Mrs. Emily Winthrop Miles, 1946*

cals and processed so that it became a skeleton of metal oxides which glowed brightly when suspended over a Bunsen burner flame. Both devices are used in camping lamps today. At the time, such inventions made gas lighting easier and more efficient—and gas illumination became more acceptable.

It was not until 1900 that incandescent gas lighting became established. Until then, gas could only be transmitted to sites relatively close to its source. Bottled gas made greater dissemination of gas lamps possible. Many bottled gases were developed, but the most successful by far was that invented by Julius Pintsch of Germany. It was used in Europe and the United States for many years, particularly to light up railroad cars, steamships, and lighthouses. By 1910, ninety-three

Pintsch gasworks dotted Canada, the United States, and Mexico. The threat of electric lamps soon became real, however.

Incandescent electric lamps underwent an evolutionary discovery process. Contrary to general understanding, Thomas Edison did not invent the electric lamp or the light bulb; he perfected them. With the benefit of scientists before him, Edison patented the first *practical* incandescent carbon-filament lamp, in 1880.

Before that—and as early as 1650—Otto von Guericke of Germany discovered that light could be produced by electricity, or electrical excitation, in addition to flame. In his experiment, he demonstrated that a globe of sulfur will emit a luminous glow if rotated rapidly to produce friction. Francis Hauksbee, an Englishman, took this one step further and evacuated air from the globe. In 1802, Sir Humphry Davy placed strips of platinum in an evacuated globe and demonstrated that metals could be heated to incandescence electrically to emit light for a sustained period of time. To heat the metal elements, Davy created a battery containing two thousand cells. He passed a current between two charcoal sticks that were four inches apart and obtained a brilliant arch-shaped flame, thereafter named the *arc lamp*. Throughout the rest of the nineteenth century, development of the arc lamp and the incandescent lamp paralleled each other.

In another attempt to create a practical electric bulb, Warren de la Rue, in 1820, enclosed a platinum coil in an evacuated glass tube. Further attempts seemed to prove that the incandescent globe or tube was very expensive to operate and therefore not competitive with gas lighting. In 1841, Frederick de Molyeyns of Cheltenham, England, patented a new type of incandescent lamp. He placed two platinum wires or filaments, bridged by powdered charcoal, within an evacuated glass sphere. The passage of current through the filaments heated the *charcoal* to incandescence. This process turned the lamp black, however, and the globe became too darkly colored to emit much light.

Another person who achieved some success in perfecting incandescent lamps was Sir Joseph Wilson Swan, an English physicist, who, in 1850, devised carbon filaments made out of paper. He later used cotton threads treated with sulfuric acid and mounted them in glass vacuum bulbs. (The vacuum process was invented in 1875.)

Swan and Edison worked concurrently toward the development of efficient, effective lamps, and both made use of a vacuum tube. Edison, however, spent more time developing dynamos and other equipment necessary for multiple circuits. He proved to be more farsighted.

On October 21, 1879, Thomas A. Edison lighted a lamp that contained a carbonized thread for a filament. The lamp burned steadily for two days. Continuing his experiments, Edison was able to extend the life of his lamps to several hundred hours. The first incandescent lighting

Early-twentieth-century American, made by Handel Co. The glass is patinated bronze in the Art Nouveau style (16 3/4" high, 16 1/2" wide). *Courtesy: The Metropolitan Museum of Art, Gift of Mrs. H. S. Mesick, 1962*

system was installed at Edison's laboratory in Menlo Park, New Jersey, on December 21, 1879. The next commercial installation, in 1880, consisted of 115 lamps on the steamship *Columbia*. The system lasted for fifteen years. More than 150 installations followed in the next two years, and in 1882 a generating station was set up in New York City. The incandescent electric light was on its way.

The search went on for more efficient filaments for incandescent lighting until Alexander Just and Franz Hanaman of Vienna learned how to use tungsten. By 1910, William D. Coolidge of Schenectady, New York, discovered a process for producing drawn tungsten filaments and greatly improved the durability of tungsten lamps. Other refinements were also explored: the use of inert gases; better sealing techniques; frosting the inside of the bulb; and using coiled filaments.

Today, in addition to incandescent lamps, there are many kinds of man-made lighting sources—arc lamps are used now to light large areas; there are gaseous vapor lamps which employ mercury; and other gaseous-vapor-discharge lamps, one of which is the fluorescent lamp.

LIGHT AND DESIGN—PAST AND PRESENT

These lamps are glass tubes, coated on the inside with a phosphor and filled with mercury vapor and a small amount of argon (which aids in starting). The ultraviolet radiated from the charged mercury gas excites the phosphor which in turn radiates within the tube. The fluorescent lamp delivers more light per input of power—nearly three times that of an incandescent lamp. Other gaseous-vapor-discharge lamps we are familiar with are flash bulbs, black-light lamps, sunlight lamps, neon lamps, and a new generation of incandescents that are filled with krypton or other gases.

The history of lighting tracks human progress. It took millions of years for humans to learn how to cope with fire and turn it into portable light, thousands of years more to develop methods for using oils for lamps, hundreds more for the development of the candle, gas lamps, and finally the electric lamp. Now that man has learned to control light, we are more dependent upon it than ever.

A lamp of today in a vinyl cocoon sprayed over an armature. By Atelier International, Ltd. *Courtesy: Atelier International, Ltd.*

On the left, a lamp of painted metal with a leather beanbag base, 18" high. Designed by Gino Sarfatti, 1966. Manufactured by Arteluce, Italy. On the right, a lamp with a painted metal housing, 20" high, designed by Gregotti, Meneghetti & Stoppino for Arteluce, Italy. *Collection, The Museum of Modern Art, New York, (left) Gift of Bonniers, New York, (right) Gift of the manufacturer*

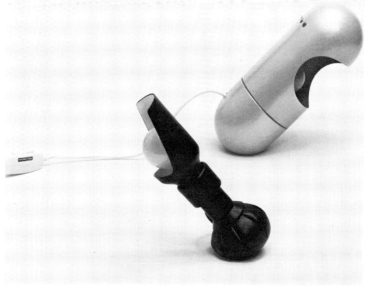

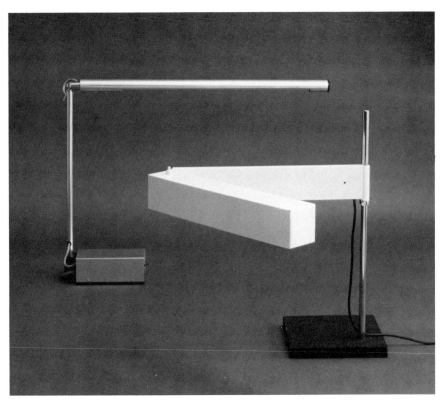

On the left, a desk lamp by Gerald Abramowitz made of anodized aluminum, 20" high, 1962. Manufactured by Best & Lloyd, Ltd., England. On the right, a desk lamp by D. Waeckerlin, of painted metal, 20" high. Manufactured by Baltensweilder, Switzerland. *Collection, The Museum of Modern Art, New York, (left) Gift of Greta Daniel Fund, (right) Gift of Design Research, New York*

A Plexiglas lamp which transmits light through the base via the edges of the form. Designed by Joe C. Columbo and his brother Gianni in 1963 for the firm of O-LUCE of Milan. The light source is a slim fluorescent by Sylvania. *Courtesy: The Late Joe C. Columbo*

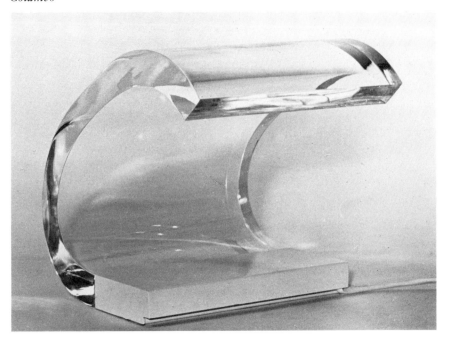

Candelabra have never really gone out of style . . .

. . . nor has the candle hurricane lamp for outdoor use. Called "Candle House" by the Rochester Folk Art Guild. *Courtesy: The Rochester Folk Art Guild*

PHYSICAL ATTRIBUTES OF LIGHT

Light is impossible to define—in spite of the fact that scientists have been trying to for three thousand years.

Early Greek philosophers thought that objects became visible because our eyes sent out special rays that bounced off objects, reflected back to our eyes, and enabled us to see the form. Today we know that, rather than sending out rays, our eyes receive rays reflected from or emitted by objects.

What we do not know is what those rays consist of. In 1665, Sir Isaac Newton hypothesized that light rays consisted of corpuscles or bullets. According to Newton, light was billions and trillions of invisible particles shot out of *luminous* bodies (objects that produce or emit light) like invisible bullets from a machine gun.

A Dutch contemporary of Newton's, Christian Huygens, developed a more sophisticated theory. He believed that light consisted of waves—and that every wave front was broken down into varying wavelengths, which accounted for our seeing colors. His theory evolved into a statement about electromagnetic energy and wavelengths, providing a more quantitative evaluation of perception.

More recently, Albert Einstein suggested that light is made up of fine bundles of energy, which he dubbed *photons*. According to his theory, photons travel with enormous velocity from luminous bodies. Though it sounds similar to Newton's corpuscle theory, Einstein's system is substantially more complex since it accounts for energy velocity and mass as well.

Even if we cannot define light properly, we do know something about how it acts. There are, for example, different wavelengths of light—some of which we can see and some we cannot. The wavelengths determine which colors we see—and the texture and brilliance of surface is also determined by the amount of light rays absorbed and reflected. Beyond that, however, we have learned some important lessons about how light functions in a living environment, and we have learned how to modulate and control that light as well.

DESIGNING WITH LIGHT

There are no strong traditions that need to be followed when designing a lamp, only needs. Because the electric light has had a relatively short history, effective ways of mounting a bulb, shielding and diffusing its light had not been developed before the twentieth century (although old lighting forms were adapted). With the advent of new materials such as plastics and the different life-styles of today, a new generation of lighting solutions is with us.

When designing lamps for interiors there are several concerns. Functional aspects of lighting such as illumination levels, how colors are affected by certain colors of light, visual comfort, and how to achieve it, as well as basic modes of lighting, are essential considerations. These are the invisibles, the solutions in designing that are not obvious when looking at the finished products in their milieu.

Another group of concerns is those that we reflect upon when purchasing a lamp—how attractive it is; aesthetic considerations and suitability of materials; practicability and simplicity of maintenance; and economy. When lamp selections are made we think about how the lamp will look in its environment, but we often neglect to consider a significant aspect—what kind of mood will be created by light from these lamps.

Illumination Levels and Visual Comfort

Consider that, once created, light travels onward from its source until it is modified in some way by striking whatever is in its path. Light, then, can glow, create shadows, model forms, reflect from polished surfaces, pass through translucent and transparent forms, distort shapes, be mystical and dramatic—depending upon how light is angled, diffused, and positioned. The angle of light changes the dynamic power of light—such as light from below, or light from above. All of us have seen the dramatic modeling of Frankenstein's monster's features through use of intense lighting from below.

There are four factors that affect how well we can see objects in an environment: contrast between object and background; reflectance of the object; its size or dimensions; and the length of time available to view an object. All these factors can be controlled. Time is irrelevant because it can be assumed that we will spend enough time in an area to see what is there.

Contrast, for our purposes, involves distinctions between colors and textures. For fine work, you often want sharp contrast so that object and background will not blend together. To create a relaxing mood the opposite may be desired.

Reflectance and absorption relate to the surface quality of objects in an environment—color and texture again. Matte surfaces and very dark objects absorb most of the wavelengths of light that reach them. Dark, matte-surfaced rooms will require more light than the same space in lighter surfaces. Something to remember is that a surface which reflects less than 10 percent of the light that hits it will seem black—no matter what its color. Different surfaces reflect and absorb different wavelengths in different degrees. This selectivity determines what color we finally see. A surface that reflects primarily long wavelengths of light (when illuminated by white light) will tend to appear red because that is the wavelength that reaches our eyes. Likewise, if most of the wavelengths reflect back to our eyes, we see white. Too much light directed at glossy surfaces—like mirrors, glass tables, photographs—will produce glare. Glare is irritating to the eyes; it interferes with concentration and will induce fatigue. To avoid it, use diffuse lighting in large areas with glossy surfaces, and apply focused, directed lighting when reading or other close work is being performed. Unshaded, undiffused incandescent or fluorescent lamps are far too bright for comfort when directed at the eyes.

In residential lighting, two concerns dominate determination of illumination levels and visual comfort: function and mood of the environment. Specific tasks, such as sewing, cooking, and reading, require higher levels of lighting than do relaxation activities. Kitchens should have well-distributed overall lighting, free from shadow.

Although illumination specialists discuss lighting levels in terms of foot candles or lumens, these types of measurements are impractical for measuring light in a household. We are more familiar with lighting output denoted in terms of wattage—which is the electrical power of an incandescent or a fluorescent lamp. Generally, a minimum lighting level for a living room measuring sixteen feet by twenty feet, or 320 square feet, would require an overall wattage of 640 watts. This could be delivered by three lamps that provide 150 watts each or 450 together, and two others using 100-watt bulbs yielding together 200 watts. This translates into two watts per square foot or twenty watts per square meter of floor area. Use this as a rule of thumb measure, but it does not take into consideration the decorative lighting that is available. Decorative lighting often provides the added amount of illumination necessary to create a mood.

Lighting should be flexible. Take into account the various activities that will take place in a room: reading, game playing, TV watching, sewing, conversation, cocktail sipping. Think about your personal visual comfort. Overhead illumination may provide minimum level lighting requirements and general lighting distribution. Decorative lighting can meet specific lighting needs—over a chair or in a nook—independent of overall needs. Decorative lighting can also impart an atmospheric effect. It is unnecessary to turn on all the lamps in a room at one time.

In a room where there are very specific activities, lighting should be designed to accommodate them. For example, in a dining room there should be direct light over the table; in a bathroom light is necessary at the mirrors; kitchen lighting should be shadowless and evenly distributed, and there should also be extra lighting at the food preparation center. Lighting should also be focused toward the point of need, such as over the pillows of beds via wall lamps or night table lamps.

Basic Modes of Lighting

The lamp's raison d'être, apart from the important decorative component, is to hold the light bulb and provide a connection to electrical current. Lamps are also necessary to diffuse light. The essential structure, then, is the same for all kinds of lamps. A bulb screws into a socket and the socket is attached to an electric light cord that in turn plugs into a socket. One can light an area with just three parts—bulb, socket, and electric wiring. But, to make for a better-functioning lamp (that isn't hooked into a ceiling or wall outlet), an on-off switching device is useful, otherwise the plug would have to be plugged in or removed from the outlet every time one wanted to control the light. In order to alleviate glare, some diffusing device is necessary. Some of these are *shades*, mounted directly on the bulb or attached to a harp that screws in under the socket; *diffusing screens* mounted at the opening of a container designed to hold the bulb, or made up of a translucent material that actually acts as a container and diffusing screen all at once; *louvers* or *baffles*, much like venetian blinds, can be constructed over light bulb containers; or light-controlling materials such as translucent glass or acrylic. Then, there are many ways of angling and positioning lamps. These range from ceiling-mounted devices, wall-mounted lamps, table lamps, floor lamps, and novelty lamps such as spotlights and light-containing sculptures.

Among these basic types of lamps there are various modes which are determined by how light is delivered or diffused. Lighting can be direct, focus downward on a particular area in one direction, or indirect, focusing upward. Or a lamp can provide semidirect light with the majority of light focusing downward and some light diffused through the lamp container. Similarly, a lamp can transmit semi-indirect lighting with the result that most of the light shines upward and some light is diffused downward. The fifth type is uniform lighting which is projected from a lamp container so that light is evenly diffused in all directions.

The kind of light bulb or tube also affects the quality and quantity of light emitted from a lamp. Curiously, the quality of light we want is usually governed by our past experiences. We expect the color of illumination to look the same as the light we have seen coming from the combustion of gas, oil, or candle wax. We also expect the shape and color of bulbs to resemble those that Thomas A. Edison standardized. Only recently have new shapes been readily available. (See the diagram of lamp shapes.)

Even though we know that color looks different in daylight from how it looks under lamplight, we have gotten used to distortion of color by incandescent lamps. Yet, there are many choices available to us here, too, as indicated in the Lamp Selection chart on page 22.

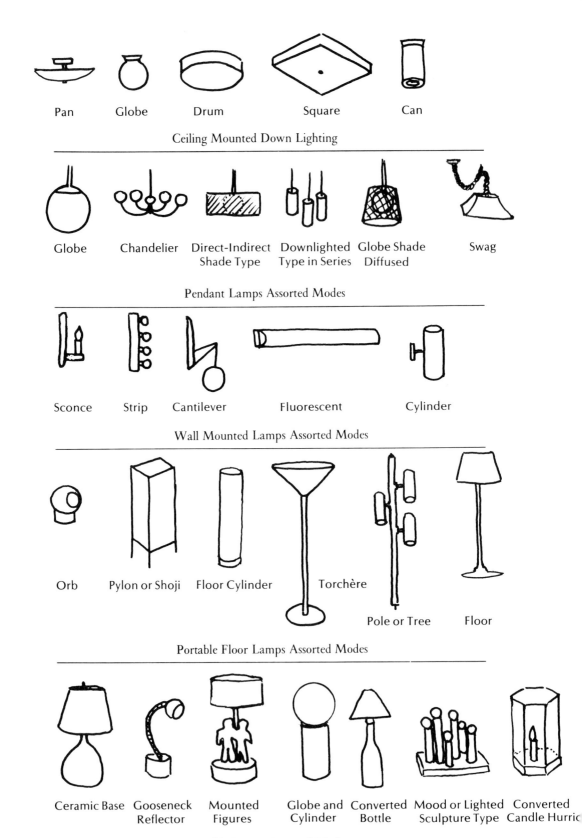

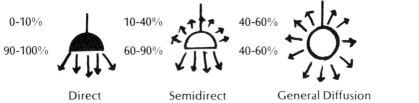

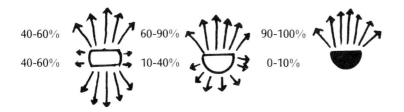

Upward and Downward Distribution of Light in Various Lighting Modes

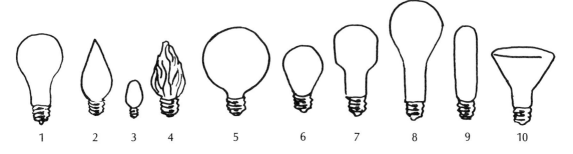

Some Typical Incandescent Bulb Shapes

1- Standard Bulb
2- Flame Shape
3- Christmas Light of Night-Light
4- Flame Shape
5- Round or Globular
6- Small Pear Shape
7- Krypton-Filled Long Life
8- Pear Shape Three-Way
9- Tubular
10- Reflector Type

Decorative Bulb by Durolite Corp.

Some lamps tend to "flatter" by emphasizing the dominant color while subduing complementary colors, others tend to provide a crisper atmosphere, bringing out true colors, as in daylight. Colors of the environment, the kind of room being lighted, as well as varieties of activities going on in the room, all should be considered when selecting a bulb or tube. For more on moods created by colors in lamps, see the section that follows on "Mood-Creating Aspects" at the end of this chapter.

Appearance

These are aesthetic aspects of design. Needless to say, the range of possibility in lamp design is huge. Not only do we have stylistic influences of past periods, such as "Early American," "Victorian," "The Twenties," "Thirties," and so on, but we have innovations of our own period. The advent of various kinds of plastics alone opened up a plethora of possibilities.

There are some general principles that apply to all lamps of all periods:

Materials used should not be imitators, but look like what they are. Plastics, for instance, should not be made to imitate wood grain or fabric textures. Each kind of plastic has its own intrinsic quality. If a wood grain is desirable, then wood should be employed.

The lampshade should not compete with the lamp base (and the converse) but complement it. Too much busyness distracts from what might be an attractive base or shade.

Lamps should suit the furnishings of a room. Simple contours fit best in contemporary settings. More elaborate designs would relate to period furnishings. It must be recalled, however, that there are just a few adaptable traditions in lighting since the electric bulb is a comparatively recent invention. It is possible to electrify antiques (without destroying them) and add a lampshade.

Lamps should be in proportion to the scale of furnishings, height of ceilings, and size of the room. Looking at extremes, a large dominant lamp has no place in a tiny room, nor would small pendant lamps fit areas that are not intimate in size or where furniture is large in scale.

Shade and base should relate and not overpower each other. It is not easy to state a general rule of proportion here because so many factors are involved. What is desirable is a balanced effect. One just has to try out shade on base and judge visually what looks right.

Practicability and Simplicity of Maintenance

How useful will a lamp be for its specific purpose? Would it be best to make a pole lamp so that multiple lights would reflect on various activi-

Gas or kerosene lamp style converted to electricity. Pewter-finish table lamp with a dark green shade and a glass chimney, 24" high. *Courtesy: Keystone Lamps*

The influence of tradition. Pewter candle-holder type design, converted to electricity, 17" high. *Courtesy: Keystone Lamps*

The texture, size, and color of the lampshade and base should relate to their environment. In doing so, an ambience is created that highlights the setting. A hole was drilled into a royal blue Mexican glass vase to accommodate the cord.

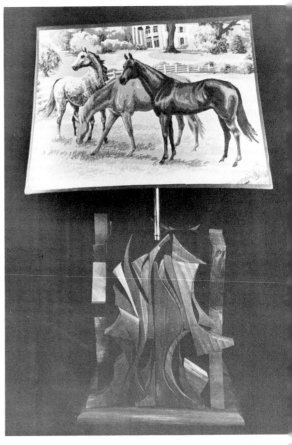

Here, although the lampshade is in good proportion to the base, the two themes of base and shade compete for attention and *do not* complement one another.

ties? Should a lamp design become a focus of attention or would it look best in an unobtrusive material or design that will blend in with the furnishings? Will materials hold up or will they soil easily, fade, discolor, or crumble? Remember that an incandescent lamp bulb generates heat. Can the bulb easily be replaced? If wiring goes bad, can the lamp be disassembled easily for repairs?

These are just a few questions that imply preplanning of design and conscious thought to the functions of the lamp and where it will be placed. Certainly, parts should not be glued permanently if that will impede the disassembling process. Bulbs, switches, plugs, sockets, and wiring do wear out eventually. Unless you are willing to scrap the entire lamp, plan the structure so that rewiring can be accomplished easily.

Suitability of Materials

Suitability of materials requires good sense and care in selecting what is to be used. Types of materials employed relate not only to the style of furnishing but to the basic design of the lamp. For example, the choice of soft fabric for shades belongs more to traditional lamps than to contemporary, where rigid, washable shade coverings are more prevalent.

Most importantly, whether a lamp will last depends upon the proper selection of materials. Certainly, Styrofoam should not be used on a lamp base where a heavy, weighted material such as marble would provide greater stability. And a plastic with a low melting point should not be near or in contact with a high wattage bulb that generates much heat. The plastic material might discolor, distort, sag, or melt completely out of shape. Nor should a high wattage bulb be enclosed in a small space covered with a fragile glass. Vents should be provided so heat can rise out of the container.

Fabric shades should not be used in bathrooms and kitchens where steam, smoke, and grease will quickly discolor the material.

Economy

We can think of economy in two ways: choice of materials and ways to save electricity.

If we put a lot of work into making a lamp and are proud of our creation, then we will want it to last. Use the best findings for constructing the lamp and making a shade. For example, solid brass usually looks better over the long run than brass-plated parts. There is, however, nothing wrong with making use of found objects. "Found" implies only that you have thought of another use for a material. Certainly, be concerned about permanence and performance. A wax-coated milk container would not work as a lampshade or base, but an empty wine bottle can easily be converted to an attractive, practical lamp.

There also are some ways to design electricity-saving factors into lamps. Some sockets and cord switches come equipped with dimmers. When a lower level of lighting can be used, the switch can be adjusted to dim the lamps, thus cutting down power usage.

Try substituting lower watt bulbs in hallways, foyers, basements, and garages.

Krypton-filled incandescent lamps use 8 percent less energy for the same performance than conventional "extended service" bulbs. The 55-, 92-, and 138-watt sizes match light output by the 60-, 100-, and 150-watt "extended service" bulbs.

A 40-watt fluorescent bulb gives more light than a 150-watt incandescent, and fluorescents last up to thirty times longer.

A fluomeric bulb, combining gas discharge, incandescent, and fluorescent in a bulb that screws into the normal socket lasts ten to fifteen times longer than most incandescents. Bulb wattages of 160 to 1250 substitute for incandescent lamps of 200 to 1500 watts respectively. These bulbs are chiefly used in industry, but have application where a single lamp is needed to light a large area.

Use reflector lamps when you want to light up a specific task area. For example, reading in bed can be done satisfactorily using a 30-watt reflector bulb instead of a 100-watt night-table lamp. Reflector bulbs also are designed to spread their beams more widely or narrowly, depending upon the need.

Mood-Creating Aspects

Skillful application of lighting can, more than any other element in the architectural environment, affect the experience of the viewer. Light is an element of design which should be used not only for visual comfort, but also to achieve predetermined emotional responses from the lighted environment. Through use of lighting patterns of varying levels of illumination, and of color in the light source and in the illuminated object, it is possible to produce certain moods such as: solemnity, restfulness, gaiety, activity, warmth, and coolness. The lamps themselves can be used to dramatize elements of interior design—line, form, color, pattern, and texture.

LAMP SELECTION: TYPE AND COLOR

Lamp Names	FLUORESCENT				INCANDESCENT	
Type	Cool White	Warm White	Daylight	Soft White/Natural	Filament-Standard	Gas-filled
Efficacy	High	High	Medium-High	Medium	Low	Medium
Lamp Appearance on neutral surfaces such as white walls	White	Yellowish white	Bluish white	Purplish white	Yellowish white	White to yellowish white
Effect on "atmosphere" of room	Neutral to moderately cool	Warm	Very cool	Warm, pinkish	Warm	Cool
Colors intensified	Orange, yellow, blue	Orange, yellow,	Green, blue	Red, orange	Red, orange, yellow	Orange, yellow, blue
Colors toned down or grayed	Red	Red, green, blue	Red, orange	Green, blue	Blue	Red
Notes	Blends with natural daylight	Blends with incandescent filament light	Similar to cool white	Tinted effect	Color rendering we are used to seeing	Blends with natural daylight

Adapted from General Electric Co. material

Higher levels of lighting generally produce cheerful effects and stimulate people to alertness and activity, whereas lower levels tend to create an atmosphere of relaxation, intimacy, and restfulness.

Lighting also can be "soft" or "hard." Soft lighting is diffused light that minimizes harsh shadows and provides a more relaxing and less visually compelling atmosphere. Too much diffused lighting lacks in interest and can make a room dull looking. Careful use of "hard" or direct lighting can provide highlights and shadows, model form, and emphasize texture. Usually, direct lighting is not used as a primary source of illumination, but rather as a supplement. The shine of metal, reflections of crystal, rich texture of other materials, create a sense of aliveness in an environment.

Colored lights can enhance an atmosphere. Certainly colored lights are heavily used in stage sets to create moods. In a home, however, they should be applied with much restraint—and be much less intense or saturated. When overdone, colored lights destroy the appearance of materials and people's coloring. A case in point: rosy-tinted lamps can flatter, but red lights wash out redness of lips and cheeks and undo the flattering effect, as can be noted in some nightclubs.

To design a more solemn atmosphere, try using subdued patterns of light but do emphasize dramatic points in the room to avoid a monotonous effect. Use color sparingly.

To achieve a restful effect, use low brightness patterns, no visible light sources, subdued color, dark upper ceiling, low wall brightness.

By way of contrast, to develop a sense of alertness and activity, employ high levels of illumination with lighting focused over specific tasks or areas.

For imparting a sense of warmth, use colors at the red end of the spectrum—pink, orange, amber, yellow. And for coolness, use colors at the cool end such as violet, blue, and green. Be aware that blue, blue green, and green detract from the human complexion and, when used indiscriminately, produce ghastly effects.

For a mood of gaiety, utilize higher levels of illumination, perhaps with kinetic lighting elements.

Positioning of lights so that rhythms are created by light and shadow, reflections and diffusions, can impart an active dynamic effect to the atmosphere.

One cannot, when dealing with light and thinking about mood effects of light, discount its relationship to color. Light and color are basic human needs. Both are necessary for sound mental health. Deprivation can be harmful.

Light also is life-giving basic energy for any organic existence. And it affects the rhythmic processes of life—our biological clocks. Man has evolved to perform under eight hours of daylight and sixteen hours of incidental light and near-darkness. So the room to which we retire, be it bedroom or living room/cum bedroom, should have built into the lamp designs lights that emit low levels of illumination. This produces a quiet mood. One extreme experience comes to mind. After leaving the clattering, noisy subway and passing through a doorway that leads directly from the subway to Rockefeller Center, one enters a black marble area where there is dim lighting. The immediate effect is startling because people tend to suddenly lower their voices to a whisper after competing at shouting levels a moment before. This effect was designed into the environment through the mood created by light and color that induces a hushed, quiet atmosphere.

By way of contrast, to develop a celebrating mood, or a sunny, joyous mood, the sparkle of reflections from objects and bright lights helps to make light of an environment. On the other hand, for a romantic environment, one would employ low levels of lights, tinted on the warm side so that light flatters as it modulates our face and form, much like the feeling one gets when entering a candlelighted room.

Positioning of lights affects the modeling of forms. Side lighting intensifies modeling; front lighting flattens it. Light affects not only the surface and structure, but the ambience.

2

Constructing and Wiring the Basic Lamp

Lampmaking requires only a few basic skills: the fundamentals of electrical wiring must be mastered, and the lamp craftsman must become familiar with available lamp parts and with elementary construction techniques. All these requirements are easily fulfilled. Electrical connections are straightforward and require few tools beyond a wire stripper and a screwdriver. Lamp construction is quickly learned and parts are readily available.

There are important construction considerations germane to each type of lamp. Table lamps should be firmly constructed—and should never be top heavy. Findings which will properly support the structure must be soundly combined, and an accessible on-off switch provided for, too.

Floor lamps involve similar considerations. Because floor lamps are often on a larger scale than table fixtures, one must consider the sturdiness of the materials. For example, rather than using a thin piece of pipe in a floor lamp, one might opt for thicker tubing for extra strength. One should also be aware of the placement of wires. Floor lamp cords should not threaten to trip pedestrians.

When the lamp will hang from a wall or ceiling, the stress created must be accounted for in construction. Stronger findings may be called for, and lamp weight should be minimized to whatever extent possible. Wiring may be different for the wall or ceiling lamp as well. While most table lamps are completely portable and may be plugged into any socket in the house, ceiling lamps often hook up to wiring inside the ceiling. Wall lamps may fit directly into the wall, or they too may plug into an electrical outlet. Wiring elements are discussed below, but there are practical considerations. Neither ceiling nor wall lamps should be constructed so as to interfere with free movement in the spaces below or around them. If wires extend from the ceiling or wall fixture, care should be taken to make them unobtrusive and keep them out of trafficked spaces.

Many of the same elemental considerations apply in the construction of novelty lamps. In designing and constructing any lamp, one must ask 1) Does it work? 2) Is it practical in the particular application (or is it serving its decorative function)? 3) Does it pose any hazard (due to positioning, wiring, or construction)? If the final result is solidly constructed, properly weighted, neatly executed, and produces the desired lighting effect, the lamp meets basic technical demands. The aesthetic aspects are entirely up to the individual. The rule of thumb is: "If it works, it's good."

There is one other rule which has no caveat: electrical connections must be planned for in advance, and they must be carefully executed. This is not meant to sound ominous—it is only meant to state the case. Wiring should be the one area immune to shortcuts. There are a few fundamental physical principles which determine what is acceptable and safe when working with electricity. Deviation from standard procedure in lamp wiring can be hazardous. So take your time, make all splices and connections with care, and do it by the book.

LAMP PARTS

Lamps may be constructed from found objects, odd findings, parts from old lamps, random pieces of hardware, and scrap pieces of wood, metal, and plastic. But for consistent, professional results, it is difficult to surpass the array of lamp parts designed specifically for this application by Angelo's LAMPARTS, Leviton, and other specialized manufacturers. Lamp parts, moreover, are readily available through hardware stores, lamp shops, and even some department stores. The advantage of using specific parts lies in the extended range and the ease of combining elements that are designed with compatibility in mind.

Because lamp craftsmen have been practicing the craft for such a long time, we have all benefited. There are parts for every use. If an ugly nut would be exposed, decorative brass washers (check rings) will cover it up. If you want to use two bulbs in a single socket, an adapter is available for that very purpose. The list of convenience parts is endless, but, even so, most lamps are constructed very similarly, and most employ only the most basic parts. The trick to lamp construction is not to be bound by the cleverness and gadgetry, but to stick to basics, and remain flexible in conceiving lighting ideas. The object is to develop lighting solutions that provide light and are structurally stable. Usually, you will discover more than one alternative in constructing any lamp. With those thoughts in mind, we turn to a catalog of the parts employed most frequently in making lamps.

PARTS COMMON TO MOST TYPES OF LAMPS

Lamp pipe, all-threaded steel pipe in standardized sizes, is the basic structural unit in most floor and table lamps. Smaller pieces are used in most other types of lamps as well. Most often, a piece of lamp pipe runs the entire length of a lamp—from socket to base—and acts as a steel backbone. Although different sizes of steel pipe are available, the standardized size, which screws into most threaded lamp parts and fits a broad range of washers and nuts, is referred to as 1/8-inch I.P.* It has an inner diameter of 1/8 inch, and an outer diameter (O.D.) of 3/8 inch. This pipe, in addition to providing excellent structural support, allows for the passage of electrical wires through its hollow center and enables the craftsman to suspend the socket and shade above the body of the lamp.

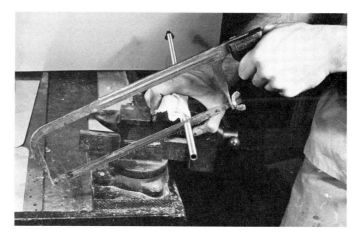

Threaded steel pipe (1/8-IP) can be cut to any length using a hacksaw. To protect the threads, cloth is wrapped around the pipe before inserting it in the vise. Before making a cut, spin two locknuts onto the pipe across the area where the cut is to be made. After the cut, slowly "back out" the locknuts over the cut threads. This serves to straighten any threads which were bent or damaged by the cutting. If any burrs remain, file gently.

*I.P.: plumbing trade term for an iron pipe with a 3/8-inch outer diameter.

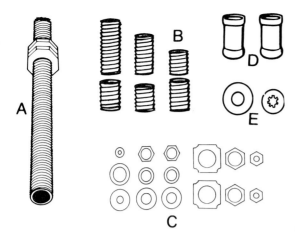

The "backbone" structure of most lamps can be created using these findings: A) threaded steel pipe; B) shorter pipe lengths called steel nipples; C) an assortment of round, square, and hexagonal nuts; D) couplings for connecting two pieces of pipe; E) slip and lock washers. *Courtesy: LAMPARTS, Angelo Brothers Co.*

Steel pipe is easily cut to necessary lengths with a hacksaw. For convenience, pipe is usually available in three foot lengths—approximately the correct size for table lamps—but longer lengths are available as well, and several lamp parts manufacturers also package short lengths of threaded pipe known as *steel nipples*. The nipples often prove to be indispensable in making connections of canopy to ceiling, candelabra socket to brass sconce, or for spacing decorative findings. Of course, nipples may be cut from longer lengths quite easily, and most craftsmen find that the added accuracy in cutting your own when you need them outweighs any convenience factor.

Steel pipe is available in its natural gray color, and a brass-plated finish has also become popular. For all its utility, however, threaded pipe is by no means decorative. To mask the pipe in areas where it would otherwise be exposed, lampmakers employ brass and brass-plated tubing of 7/16-inch I.P. This tubing slides over the steel pipe, camouflaging the threads and creating a smooth line. These tubes can also be cut to proper lengths with a hacksaw. In addition, brass tubing with 1/8-inch I.P. and 3/8-inch O.D. may be used just as the steel pipe is—to provide the basic structural support. In that role, the ends of the pipe are "tapped" or threaded so that they can accept nuts and washers.

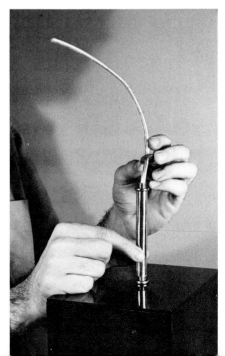

Brass pipe (7/16-inch inner diameter) slips over less attractive threaded steel pipe. Check rings are used here to hide an ugly steel locknut and to serve as decoration. A round brass nut is used to hold the check rings and brass pipe in place.

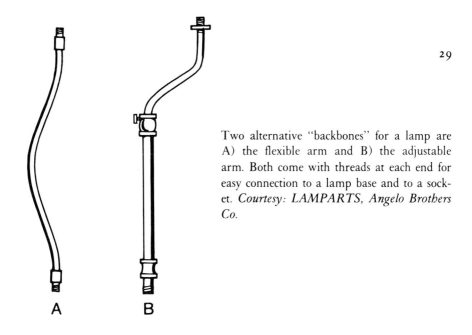

Two alternative "backbones" for a lamp are A) the flexible arm and B) the adjustable arm. Both come with threads at each end for easy connection to a lamp base and to a socket. *Courtesy: LAMPARTS, Angelo Brothers Co.*

Other piping solutions include *adjustable arms* and *flexible arms*. The adjustable arm consists of two pieces of pipe, one inside the other, and sliding is controlled by a side screw. *Flexible arms*, usually available with threaded ends for easy application, bend, twist, and curve in any direction. This material allows for the light source to be directed conveniently and quickly.

With threaded pipe or end-tapped brass pipe, *round nuts* or *hexagon nuts*, and *lock* or *slip washers* are essential. Nuts are threaded inside so that they screw around the pipe. Washers are unthreaded. They slide over the pipe and are usually placed between any surface and a nut to protect that surface and allow the nut to get a firmer grip on the threads. In lampmaking, both the nuts and washers are available in steel, brass, and aluminum. Nuts may be round, square, or hexagonal. Washers most often are round and of either steel or brass. Nut and washer combinations are employed to hold different parts of the lamp in place. At the base, for example, the cross section would look like: nut, washer, base, washer, nut. The nuts, when tightened toward each other, hold the base securely in place. The same sort of mechanical attachment is used to secure the top of the base, the harp, and the socket as well, but, in those areas which are most visible, round brass nuts (sometimes covered with *check rings*) are used, and the washers are eliminated.

In addition to the standardized size used for the majority of lamp part threads, other standard sizes are employed in special applications. A smaller thread (1/4–27) is found on lamp *harps*, for example. And larger steel pipe, with a 1/4-inch I.P. and 1/2-inch O.D., is often used in large structures—such as floor lamps, which require a heavier "backbone," and sometimes in ceiling connections.

Accessory parts are made for use in *coupling* pipes of the same dimension and *adapting* pipes of different dimensions. When pipes of the same dimension must be joined, *couplings* are used. The pipes screw in from each end, with the threaded coupling forming the joint. Adapters work in the same manner, except that one end accepts one type of pipe, while the other end accepts a different kind—either smaller or larger or of different threads.

Finally, we reach an important category of basic lamp part which has more aesthetic than practical application. *Check rings*, or *seating rings*, fit over standard steel pipe and completely cover unsightly nuts and washers. Check rings come in many sizes, but the inner diameter is usually 7/16 of an inch, a size that allows it to slip over standard steel pipe and fall over the tightened nut and washer (a veritable nutshell). Check rings generally serve ornamental functions—as plates, dividers, or as transition findings between a pipe and socket, or between a pipe and a lamp base. They add a finished, professional touch to lamp constructions.

A few important lamp findings. A) 7/16-inch inner diameter brass pipe; B) 1/8-IP threaded steel pipe (and a shorter nipple); C) two devices for plugging up the top of a bottle or vase, converting such containers into lamp bases; D) large and small check rings; E) three hexagonal nuts of varying thickness and one round brass-plated nut; F) a coupling; G) the arm of a two-piece detachable harp; H) the base of the harp into which the arms are fitted; I) a butterfly clip which clips onto a bulb and supports a lightweight shade.

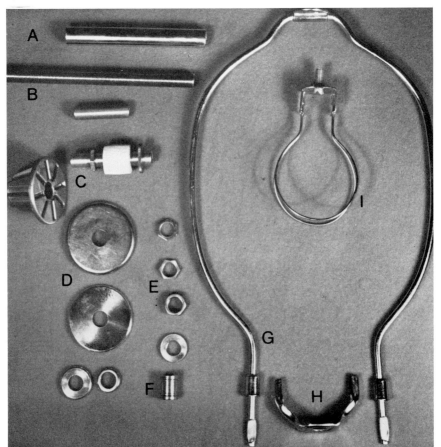

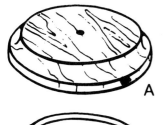

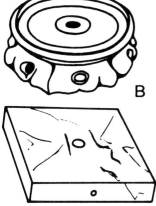

Lamp bases are available in A) unfinished hardwoods, B) cast metal, C) marble, and many other materials. Holes are drilled to allow for easy lamp assembly and wiring. *Courtesy: LAMPARTS, Angelo Brothers Co.*

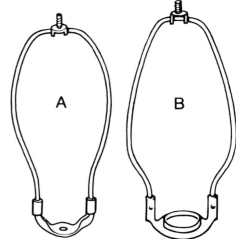

The two most commonly used harps are A) the two-piece detachable harp and B) the one-piece screw-on harp. The latter screws directly onto the top housing of a socket. *Courtesy: LAMPARTS, Angelo Brothers Co.*

TABLE LAMP PARTS

A few lamp parts are used almost exclusively in making table lamps. Most designs require a properly weighted *base*, so that the lamp will not be top-heavy and topple easily. *Lamp bases* that are naturally heavy —marble, plaster, ceramic, wood—eliminate that balance consideration. And for lighter materials, special weights are available. Lamp bases can be made of virtually any material. Commercially manufactured lamp bases are available in marble, hardwoods, cast metals, and plastics. But lampmakers just as often make their own bases with those materials, or modify commercial forms with paint, veneer, oils, varnish, wood stain. Usually, the lamp base has a hole through its center to accommodate the threaded pipe that acts as the basic structural support. The bottom of the base sports a recessed area around the hole to house the nut and washer that hold the pipe in place. And, in solid bases, a channel to house electrical wires runs from the hole to an outer edge of the base.

Any lamp that will require a shade must be constructed with that in mind. Shades are attached to table lamps in two basic ways: *harp* or *clip-on*. The clip-on shade does not require any structural planning— except to the extent that it is used only for very light shades. The *clip-on* fits directly onto the light bulb. If the shade is too heavy it will continually slip and possibly break the bulb. For heavier shades—for most shades used on table lamps—some form of *harp* is used. The harp, simply speaking, is a loop of wire which encircles the light bulb. It is generally attached to the lamp pipe stem, or to the bulb socket, and it has a threaded pin at its top. The shade fits over that pin, and a screw-on cap keeps the shade in place. There are two types of harps. One is the *two-piece detachable harp*, which fits into a bracket mounted on the lamp pipe stem. To fit the harp into place, the sides are compressed and fitted into the bracket, and, when released, the pressure keeps the harp in place. The other basic type of harp is a single-piece unit that screws onto the threaded top of the bulb socket. Unlike the bracket for the detachable harp, which is bolted onto the stem, the *single-unit harp* may be completely removed by unscrewing. Harp sizes vary in shape and height. Wider harps accommodate larger bulbs. Higher harps support taller shades. The proportions are discussed in chapter 3. Shades may also be raised—even though the harp is low—with *shade risers*. These threaded brass pieces are available in 1-, 2-, and 3-inch lengths; they screw onto the harp, and the shade fits over the riser. Shade attachment is completed with *finials* that screw onto the harp and hold the shade securely in place. Finials range in design from simple cylinders to ornate eagles. The style should fit the lamp.

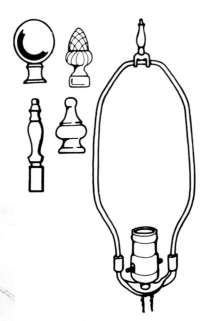

Harps, which hold shades, are topped off with finials—small decorative nuts which screw onto the swivel at the top of the harp. *Courtesy: LAMPARTS, Angelo Brothers Co.*

A common table lamp assembly employs A) a nut (hidden beneath a check ring) that tightens the base to the threaded pipe; B) brass tubing that covers the threaded pipe; C) a check ring used for decoration; D) a round, brass-plated nut to tighten the brass tubing; E) the base of a detachable harp, which gets sandwiched between the round nut and the base of the socket F), which is screwed onto the threaded pipe.

A similar assembly on a glas base uses a cap over the glas ornamental brass tubing (spir dles), and, again, the two-piec detachable harp.

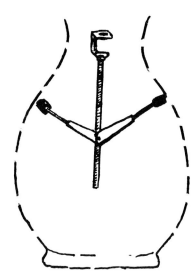

A finding called the "vase toggle" can be slipped inside a glass, ceramic, or other vase to keep the vase cap, harp, and socket firmly centered over the object. The toggle is on a thread, making it adjustable. Rubber tips protect the inside of the vase. A hickey with a 3/8-inch threaded hole comes attached (to accept a 1/8-IP nipple).

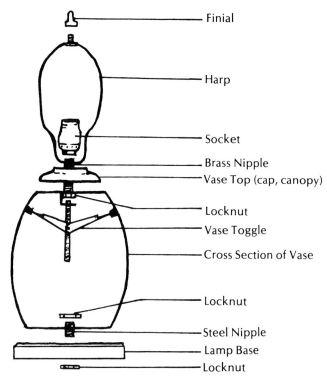

Anatomy of a typical vase lamp on a base. If the vase is not precious, a hole can be drilled at the bottom so that the electrical cord can exit there, as discussed in chapter 6.

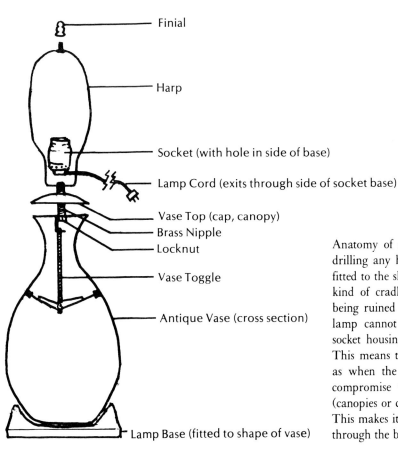

Anatomy of a lamp using an antique vase. To avoid drilling any holes in an antique, a lamp base which is fitted to the shape of the vase is employed. By using this kind of cradle, the antique sits securely, but without being ruined by holes. Note that the lamp cord in this lamp cannot pass through the inside of the vase. A socket housing must be bought which has a side hole. This means that the wire cannot be as cleverly hidden as when the vase has a hole, but it is a reasonable compromise when working with antiques. Vase tops (canopies or caps) are available with side holes, as well. This makes it possible to pass the cord out of the socket, through the brass nipple, and out through the vase top.

CEILING LAMP PARTS

The canopy is the ceiling lamp equivalent of the table lamp base. Generally made of plated or painted metal, canopies cover the space in the ceiling that houses the electrical junction box. Through a hole in the center of the canopy extends the lamp wire for a flush or hanging lamp. Canopies are attached to the ceiling junction box with a steel nipple and *crossbar*. The crossbar is a strip of metal that is screwed onto the lugs of the ceiling box. The nipple is then screwed through a hole in the crossbar's center and adjusted so that the canopy fits snugly against the ceiling, with the nipple protruding through a 7/16-inch-diameter hole in the canopy center. A lock nut screwed onto the nipple mounts the canopy against the ceiling. The standard crossbar is 4" long, but an all-purpose, universal crossbar, adjustable for 3-, 3 1/2-, and 4-inch ceiling boxes is available, as well as an offset, or recessed, version of the 4-inch bar which is necessary when the distance between junction box and canopy is small.

Canopies are of different types and they function in different ways. The type described above is meant for use with a lighting fixture that is wired directly into the house circuitry and is usually controlled by a switch on a wall. Such canopies may have a single hole, to accommodate a single wire (though many lights may be part of that fixture), or they may have many holes for several wires, each supporting one or several individual bulbs. In addition, some canopies have hooks to suspend chains that are part of the fixture. While most canopies cover a

Many ceiling fixtures call for the use of a canopy. Six varieties of canopies are shown here. At upper left is an unusual canopy which has a built-in hook and outlet. The outlet makes it easy to change lamps by just plugging and unplugging. The hook is useful for swag lamp hanging. At bottom left and center are three standard canopy designs which come in brass, chrome, and antique finishes. These canopies double as lamp bases. At top right is a glass shade holder. It has a built-in socket. Three side screws are used to hold the lip of a glass shade. At lower right is a modified canopy with a loop to facilitate installation of hanging lamps. *Courtesy: LAMPARTS, Angelo Brothers Co.*

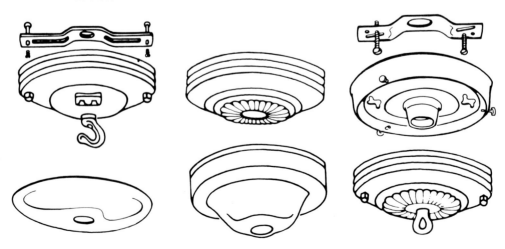

junction box, some are only plates with built-in hooks, from which fixtures are hung. Often, the lighting unit will be hung from a chain, with the electrical wire threaded through that chain. The wired lamp chain hooks onto the ceiling plate, is drawn over to and down the wall, and is plugged into a wall socket. Such units are known as *swag lamps*. Another type of canopy makes it easy to interchange hanging lamps. To avoid having to change the canopy with each lighting fixture, this canopy comes with a built-in socket for a plug. The lamp hangs from a hook and just plugs into the canopy plate with *no* elaborate wiring.

Related to the covering canopy is the *glass shade holder*. Often of brass, this type of fixture holds fitted glass shades in place with three side screws. Sometimes, too, holders are suspended from the ceiling; in that event, they are combined with standard canopies. Holders provide an excellent lighting solution for areas where unobtrusive fixtures are desired, or where space is a consideration—as in closets or hallways—since the basic unit fits directly against the ceiling.

Off the ceiling, canopies can be used as caps for gourds and for other objects being converted into lamp bases.

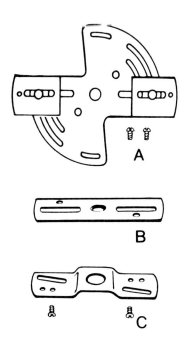

Crossbars are used to connect canopies and other ceiling fixtures to the junction box in the ceiling. The universal crossbar A) is adjustable to different size junction boxes. The standard crossbar B) fits a 4-inch box. When the distance between the junction box and canopy is too small, an offset crossbar C) should be used. The crossbars are screwed onto lugs in the junction box. Canopies are attached to the crossbars using steel nipples bolted to the center hole in the bar. *Courtesy: LAMPARTS, Angelo Brothers Co.*

If a ceiling fixture with more than a single bulb socket is desired, the twin light socket adapter will provide double the light. Courtesy: Leviton

There are other solutions to ceiling lighting. A large variety of bulb sockets (in *clusters* of two, three, or four) fit directly onto the standard ceiling socket. Clip-on and screw-on shades can be fitted over these socket units.

Dozens of other lamp parts and findings fill unique needs and extend the range of possibilities for ceiling fixtures. These include devices for adapting the center hole of a canopy into hooks or links; for extending a ceiling outlet by a few inches; for diffusing the bright light of a bare bulb.

ELECTRICAL PARTS FOR LAMPS

There are hundreds of electrical parts on the market, but the lampmaker need only be concerned with four main elements: the bulb socket, the switch, the plug, and the wire.

Bulb Sockets

Dozens of different sockets are available. Sizes range from miniature to mogul. Lampmakers most commonly use two intermediate sizes: candelabra and the standard ("medium") sockets. Socket size is one consideration in designing a lamp. Certain decorative bulbs are available only with small bases—others are larger at the base. So if a particular type of bulb is desired, be certain that the socket being installed in the lamp will accept that size bulb. For most applications, a medium-sized socket with a built-in switch serves well. Built-in switches operate by key (or knob), push-through button, or by pull chain. And some sockets have built-in dimmers which allow the user to control the brightness of the light. Dimmers represent a refinement in the control of illumination surpassing the widely used three-way bulb.

Sockets vary in finish as well as in size. Sockets are typically of brass or brass-colored aluminum, but aluminum and nickel-plated sheaths are also widely used. When the units will be hidden, porcelain or plastic or ceramic types are used. All metal-sheathed sockets are insulated with a wall of cardboard that fits between the socket itself and the outer metal sheath.

CONSTRUCTING AND WIRING THE BASIC LAMP 37

Sockets come in all shapes and sizes. A) Brass-plated socket with a push-through switch; B) clip-on socket encased in plastic, used on a string of lights; C) candelabra base socket with a hickey (which may be screwed onto threaded pipe); D) candelabra base socket with a flat base for mounting on flat surfaces; E) porcelain, standard base socket with hickey; F) two socket bases, available with brass or nickel plating; G) the socket mechanism, (*left*) with a three-way turn knob and (*right*) keyless; H) outer metal sheaths for sockets; I) standard base socket with a porcelain mounting; J) standard base socket in plastic mounting that can be screwed to a flat surface like sockets D and I; K) standard base socket in a plastic mounting with a turn knob at back.

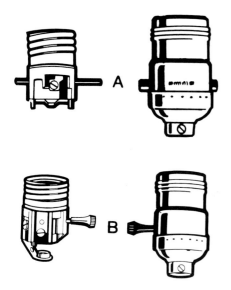

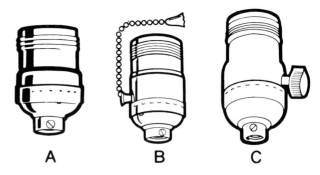

A) The socket mechanism for a push-through socket (*left*), and the same socket encased in a brass or nickel-plated housing (*right*); B) a socket mechanism with turn knob and a hickey (*left*), and a turn-knob socket that has threads and side screw in its base instead of a hickey (*right*). *Courtesy: Leviton*

A) Standard keyless socket; B) socket with pull chain; C) socket with dimmer switch. *Courtesy: Leviton*

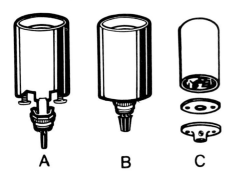

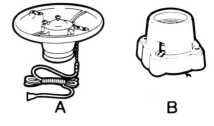

A and B) Two types of standard sockets with rotary switch bases; C) socket with a removable back for internal wiring. *Courtesy: Leviton*

A) Porcelain-encased socket with pull chain; B) plastic-encased socket which, like A, screws to flat surfaces. *Courtesy: Leviton*

Sockets are attached to lamp fixtures in two basic ways. Most have threaded holes at their bases, so that lamp pipe may be screwed into them readily. Many others, however, employ a *hickey*. Most hickeys are attached to the socket at its base. They are usually L-shaped pieces of metal, tapped at one end to accommodate threaded pipe. Their function is to extend the socket, while still permitting direct attachment of the socket to the lamp. One of the most unusual of these sockets is the *adjustable hickey*. It looks like a socket on stilts. Side screws permit extension or retraction of the socket—a versatile quality when working with candelabra lamps.

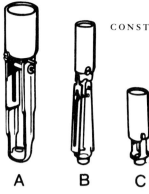

A) Standard base socket on an adjustable hickey; B) candelabra base socket on adjustable hickey; C) candelabra base socket on normal hickey. *Courtesy: LAMPARTS, Angelo Brothers Co.*

Other socket casings worthy of note are *branches* which will convert single sockets into doubles with the twist of a wrist. No wiring is required, sockets for two bulbs branch off from a single threaded male part which enters the original socket. For ceiling fixtures, it is possible to purchase clusters with two, three, or four sockets wired around a central shaft.

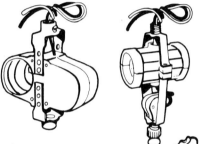
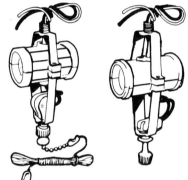

Three types of twin clusters. The one at left has no switch. The others come with a pull chain and turn-knob switch. These can be ceiling mounted or used in table or floor lamps as well. Clusters of sockets are also available with triple or quadruple groupings. *Courtesy: LAMPARTS, Angelo Brothers Co.*

Another variety of socket is designed to be screwed onto a flat surface: ceiling, wall, or base. Most often of porcelain or Bakelite, these sockets can be fastened solidly to almost any flat surface, and they are available to accommodate different lamp base sizes as well. There are, of course, many more variations on the socket theme.

Switches

The lampmaker needs only a few types of switches. As discussed above, many different types are built right into lamp sockets, but it is often inconvenient to reach for a switch so near the bulb. *In-line switches* provide a solution. In-line switches are installed along the lamp cord. They are made to be attached easily and quickly; some will dim lights, some have two or three positions of brightness, and some simply turn on and off.

In addition, *toggle*, *rotary*, *three-way rotary*, and *push button* switches may be built into lamp bases and pole lamps. Every switch essentially cuts into the live wire between the socket and the bulb, and regulates the flow of current to that bulb. More will be said about different types of switches as the wiring for each type is discussed in this chapter.

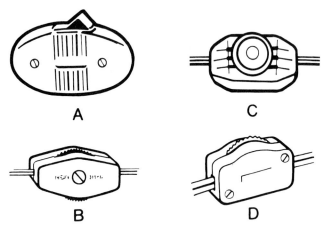

Four types of "in-line" switches. A) The standard on/off switch with screw terminals inside; B) a clip-on, thumb-wheel, on/off switch; C) a full range dimmer switch; D) a "high-low," thumb-wheel, clip-on dimmer switch. *Courtesy: Leviton*

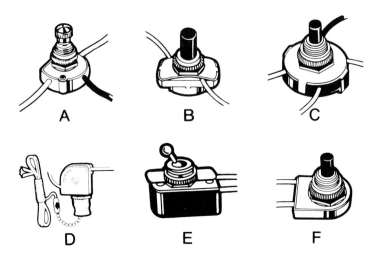

A) Two-circuit rotary switch; B) single pole rotary switch; C) fluorescent starter switch; D) pull-chain switch; E) toggle switch; F) push-button, single pole switch. *Courtesy: Leviton*

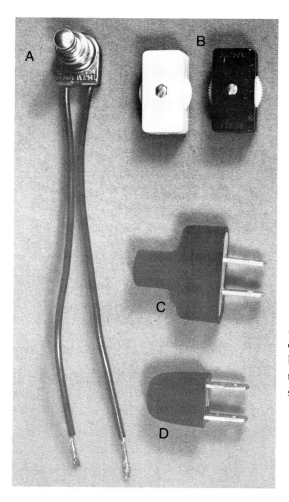

A) Single pole rotary switch; B) in-line, clip-on, thumb-wheel switches in ivory and brown; C) the standard plug with two screw terminals; D) clip-on plug that requires no stripping of insulation.

Plugs

A plug is simply a convenient and safe way of extending current from a home electrical system into a portable wire that will supply electricity to a bulb. Most plugs are made to accept the two wires commonly used in lighting fixtures. Although some plugs will have space for a third wire (a "grounded" wire), most lighting fixtures do not require this added precaution. The most commonly used plug has two prongs. These metal prongs are in contact with two screws. Both of the wires from a lamp are stripped of some insulation and connected to the plug, one to each screw terminal. A second popular variety of plug eliminates the need for separating the two wires and for removing insulation from the wires. This expeditious clip-on plug has small teeth that bite into the wire through the insulation, effecting a connection without further ado.

Wire

The primary consideration in choosing lamp wires is aesthetic. The same basic wire is used in all lamps. It is composed of fine copper filaments which are encased in insulation. Depending on the thickness of the bundle of filaments, the lamp wire is assigned a "gauge" number.

Number 18 wire is most commonly used in lampmaking. All hardware stores, lamp stores, and department stores sell #18 wire in several different insulations. The range includes: black, white, brown, gold, silver, rounded, twisted, silk-covered. The choice is completely up to you; the color and design of the lamp should be easily served by the varieties available.

ELECTRICAL WIRING TECHNIQUES

The electrical power required by a lamp is small enough that the lampmaker need only be concerned with the simplest circuitry. Most American homes operate on 120 volts in what is called a "three-wire system." The electricity flows to the house from the power source through a pair of black "live" wires. For electricity to flow, a complete circuit must be made, so the electricity which is pumped through the black wires leaves the circuit through the "grounded" white wire. The current flows in one line and out the other. (The use of two black wires makes it possible to set up more than one circuit loop, providing better distribution of current throughout the house.) The simplest lamp is wired according to this same principle: that electricity flows in one direction through a circuit. A *pair* of insulated wires are connected to every lamp socket. One is the "live" wire which feeds in current (the positive input); the other wire is a "neutral" grounded wire which provides a route for the current to exit (negative output). The socket itself sits in contact between the live and neutral wires, tapping the electrical flow.

Electrical wires are usually copper, encased in plastic, rubber, or another insulating material. Because of the chemical properties of the metal, wire is capable of propagating, or "conducting," a flow of negatively charged particles called "electrons." (The rate of this electron flow is called electrical power, measured in amperes.) The larger the wire (or bundle of wires), the more power it can supply. It is important that larger (heavier gauge) wires be used for appliances which draw a great deal of power, in order to avoid circuit overload. Most lamps create no overload problem. The amount of power they require is fairly small, and the national wiring codes set standards for the wire size required. As mentioned earlier, lamp fixtures generally employ #18 or #16 wire. Wires used elsewhere in home circuitry range from #18 to as large as #0.

As said above, there are two wires connected to any lamp socket. Note that while most lamp wire appears to be a single cord, it is actually two separate lengths of wire connected *only* by insulation. The wires are said to be "molded" together in the insulation. The two metal wires never touch. If the metals were to touch (which sometimes happens when insulation wears out, exposing the metal), it would produce a

CONSTRUCTING AND WIRING THE BASIC LAMP 43

short circuit. So it is important to remember, when working with lamp wires, to keep the wires protected from one another at all times, using electrical tape or another insulator. Repair breaks in the wires without delay.

SAFETY

In the majority of work done by lampmakers, there is no exposure to live electric current until the finished lamp is plugged into the wall socket. But some lamps—particularly ceiling and hanging fixtures—are connected directly to the house's main circuitry via lead wires built into the walls. Connecting a lamp to these leads in a ceiling box must be done with extreme care. The amperage in the average house circuit is sufficiently powerful to do bodily harm. Improper electrical work in any kind of lamp can produce heat, sparks, and/or shocks. Read the following safety points carefully before beginning to wire *any* lamp.

1. With the exception of replacing a switch or light fixture, the homeowner is required to get a permit to do electrical work, in accordance with the national electrical code and local electrical codes. It is recommended that the novice do only the elementary kinds of wiring illustrated in this text.

2. If you are not absolutely certain what you are doing in a given electrical hookup (for example, if there are more wires coming out of a junction box than you can account for), do not do it yourself. Ask an electrically inclined neighbor, friend, or a professional.

3. Know how to turn off your house's main power source by removing the fuse or turning off the circuit breakers. In case of emergency, access to the main power source may be critical. And it will also be necessary to reach this power panel when hooking up lamps directly to the house circuit.

4. If you are working with wall or ceiling wiring, first turn off the main power source. This does *not* mean just turning off the wall switch in a room. It means throwing the main circuit breaker, or removing the main fuses.

5. Do not touch any exposed wire or any piece of metal that is touching bare wire.

6. Work with dry hands, on dry ground. Do not touch electrical fixtures when you are wet or when you are standing on a wet surface. When you are wet you become a much better conductor for electricity, setting yourself up for a potentially much greater shock.

7. If a cord, lamp fixture, or switch gives off shocks or sparks, either repair or replace it immediately.

8. If paper, cloth, or other combustible materials are being used on a lamp (for instance, as a shade), be certain there is no direct contact with the bulb, and that there is sufficient ventilation to dissipate heat buildup.

9. Use wire and electrical fixtures which have the UL (Underwriter's Laboratory) seal. These parts have been tested for safety.

PREPARING THE WIRE FOR CONNECTIONS

Before two wires can be connected, some of the protective insulation must be stripped off. Most wire used in lampmaking comes molded in pairs. It is necessary, therefore, to separate the two by cutting with a knife along the midline of the insulation. Be careful not to cut through the insulation too near the metal. After splitting a few inches of the cord into two forking insulated wires, use a knife, razor blade, or wire stripper to remove about an inch of the insulating sheath from each end. If using a knife, cut into the insulation at an angle, rather than straight downward. This will reduce the probability of nicking the wire. Be careful not to cut the wire. Often the wire is not a single strand, but a bundle of thin copper filaments. If that is the case one must be particularly careful not to nick the wire when removing insulation, since a single nick may sever several of the fine strands. If you are planning to do much lampmaking, an inexpensive wire stripper is a sound investment. It comes with a series of holes that correspond to different size (gauges) of wire. By putting the right gauge wire into the correct hole, one can swiftly remove every trace of the insulating sheath without damaging the wire.

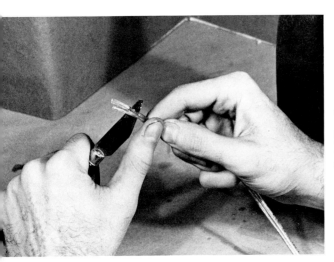

To prepare lamp wire for connection, split the cord along the midline of insulation into two forking wires. Do *not* cut through the insulation into the metal.

Using a knife or wire stripper, remove about 1/2 to 1 inch of insulation from the ends.

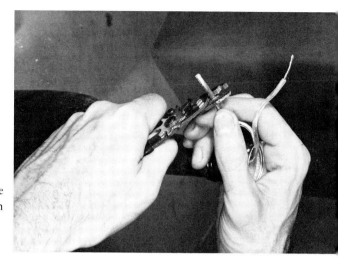

After having started to split the cord with a knife, you can easily pull the two wires farther apart, if necessary.

MAKING CONNECTIONS

Wire to Screw Terminal Connections

To connect wires to screw terminals (found in sockets and plugs), remove just enough insulation so that a loop of exposed wire will fit around each screw. Do not strip off too much insulation. Excess exposed wire can lead to shorting and concomitant electrical hazards. Generally about 1/2 inch of exposed wire will suffice. If using stranded wire, twist the exposed strands into a tight bundle to resemble a single strand. If possible, solder the twisted wires so that they will be securely fastened together, leaving no stray filaments to worry about. With a pair of pliers, bend the exposed wire into a small loop. Hook the loop around the opened screw with the open end pointing clockwise so that when the screw is tightened the wire will be drawn closer around the screw. Be certain that all strands of wire are tucked in and held under the head of the screw.

The revealed wire might actually be a bundle of fine wire filaments which should be twisted together to resemble a single wire.

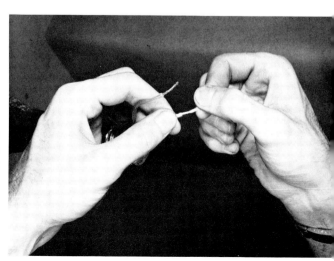

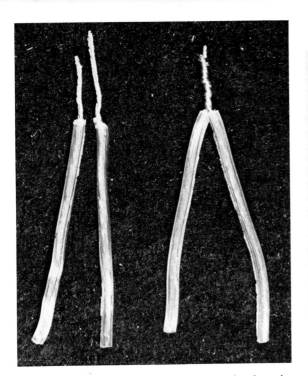

When two wires are to be connected to each other, the simplest way is with the "pigtail" splice.

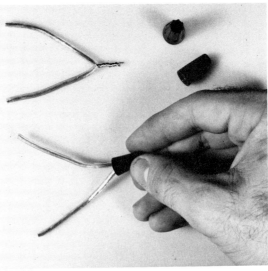

The pigtail splice can be made stronger by using solderless connectors (wire nuts) which create a safe and solid connection. Push the connector against the tip of the splice and twist clockwise until the connector grips the wire.

The Solderless Connection

Two wires can be connected one to another with "wire nuts" which screw onto the wire. Also called "insulated solderless connectors," these are best used in connections which will not be under much stress—such as between the leads of the junction box and wires from a ceiling fixture. While solderless connectors are bulkier than the connections produced by twisting or by using solder, they compensate with their convenience. About 1/2 to 3/4 inch of insulation should be removed from the wires. Twist the wires which are to be connected into a single strand. (This is a simple wire "splice" for holding two wires together temporarily.) Push the solderless connector against the tip of the wires and twist clockwise until the threads inside the connector tightly grip the wires. Make certain wires are not exposed. For extra safety, cover the entrance to the connector with electrical tape. The connectors come in three sizes; for most lamp jobs, the small and medium size connectors will suffice.

Splicing and Soldering

To obtain the soundest, strongest connection between wires, one should splice and solder. Splicing is simply a matter of wrapping the wires together. In one type, the "pigtail" splice, wires are twisted together into a single strand. Other types of useful splices are illustrated. And learning how to splice wires effectively is the key to producing a solid, soldered connection.

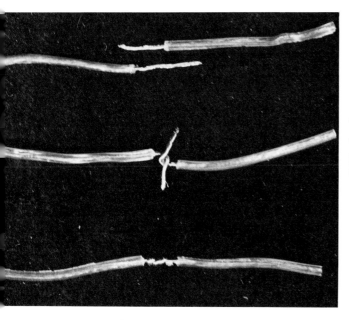

The "Western Union" splice is used to connect two wires in a straight line. At top, the two wires are positioned next to each other. They are then hooked over and wrapped around each other tightly.

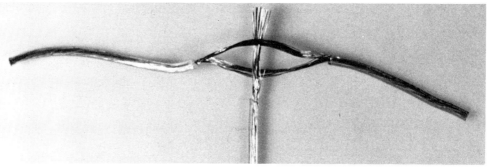

To connect a wire at right angles to another wire (as when connecting wires for "in-series" sockets), separate the copper filaments of the main wire (*horizontal*) into two roughly equal bundles. Pass the stripped end of the tapped wire between the bundles.

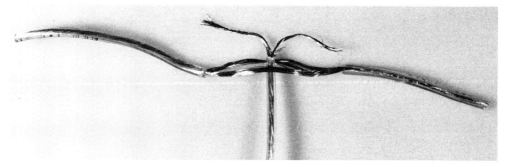

Now divide the filaments of the tapped wire into two bundles.

Wrap the two tapped wire bundles around the main wire in both directions. This forms a very strong splice.

48 CONSTRUCTING AND WIRING THE BASIC LAMP

Soldering actually improves the electrical contact between wires, in addition to increasing the physical strength of the joint. Begin by checking the quality of the splice. All straggling insulation should be removed. Wipe the wires with a clean cloth to remove grease, then sand lightly (with fine sandpaper) to give the metal some "tooth." An inexpensive soldering iron with a medium-sized tip is perfect for lamp wiring purposes. Heat the iron until it can melt the solder with ease. The best solder to use is "half and half, resin core." ("Half and half" describes the combination of tin and lead that composes the solder alloy. "Resin core" refers to the type of flux found in the core of the solder.) Apply a small amount of the solder to the tip of the iron until it is lightly coated. This process is called "tinning" the iron; it helps conduct heat from the iron to the wire. Hold the tip of the iron against the wire until the *wire itself* becomes hot enough to melt the solder. The mistake made by many beginners is to melt solder onto the iron and drip it onto the wire. This is incorrect. The *wire* is heated by the iron until the solder

To solder the spliced wires, heat the wire with the tip of a soldering iron. Touch resin core solder to the wire near the tip of the iron. When the wire gets sufficiently hot, the solder will melt.

melts from the heat of the wire itself. While holding the soldering iron's tip against the wire, touch the solder to the wire near the tip. As the solder melts onto the wire, move the solder along the splice until it evenly coats the exposed wire. Too much solder has been applied if droplets of solder form. Remove the solder from the wire. Continue to hold the soldering iron tip against the wire until the solder has flowed into any spaces between wires. Then allow the splice to cool and harden. Do not disturb the wire until the solder is hard (usually less than a

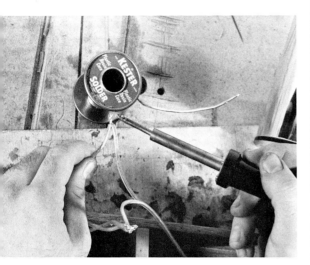

When the solder melts, spread it evenly over the splice. Then remove the soldering iron from the wire and allow to cool.

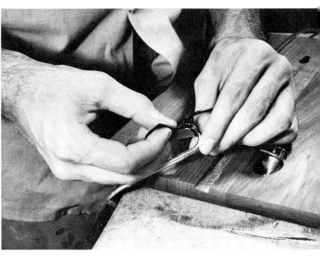

All exposed wires, such as those that have been spliced and soldered, must be reinsulated. Wrap all bare wires with vinyl plastic tape until all exposed wire is covered and the taped layer is as thick or thicker than the original insulation. Here lamp cord was spliced and soldered to a switch lead and is being wrapped with black tape.

minute). Wrap plastic electrical tape around the exposed splice. Continue to wrap back and forth over the exposed area until the thickness of the tape is approximately equal to the thickness of the original insulation.

WIRING LAMP PARTS

The Standard Plug

The standard plug requires a simple wire-to-screw terminal connection. If using molded lamp wire, split about 2 inches at the end of the cord. Remove approximately 3/4 of an inch of insulation from the two wires and twist each wire so that each forms a tight separate wire with no stray filaments. At this point, one may apply a touch of solder to the tip of each wire to hold all strands in place. Pass the wires through the center of the plug. Tie an "underwriter's knot" into the wire, as illustrated, leaving enough length of wire for each strand to reach the screw terminals. Pull the cord until the knot is tight against the plug. With pliers, make a loop in the end of each wire and screw each to its terminal. Always be certain that the exposed wires will not touch each other. Even the slightest strand touching the other terminal will short out the plug. After tightening the screws, the cardboard insulation which comes with the plug should be replaced, to protect the screws and wires.

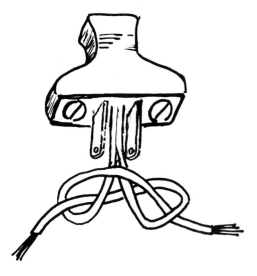

The underwriter's knot should be tied in the wires that are being attached to a plug. After tying the knot, pull the wire against the plug.

Form loops in the exposed ends of the wires and fit them around the screw terminals of the plug. Tighten the screws.

After checking to see that there are no stray wires sticking out from under the screwheads, cover the wires and screws with the cardboard insulation that comes with the plug.

Clip-on Plugs

As mentioned earlier, there are several plugs on the market that are quicker and easier to use, because they do not require any stripping, knotting, or screwing. The molded cord is inserted in a hole in the plug where it is clamped in place. Toothlike contact pins automatically bite through the insulation to complete the connection.

When plugging and unplugging lamps, never pull by the cord. Grasp the plug itself, otherwise the wires are likely to rip out and could cause a short in the process.

The Bulb Socket

Sockets use the same screw terminal connection as do plugs. Molded wires must be separated, twisted, soldered, formed into small loops, and screwed to the terminals. See the exploded diagram showing installation of a socket. The current which enters through one wire is conducted through the screw terminal to an area of contact between the socket and the incandescent bulb's metal base. From the base, the current passes through the filament and out of the bulb through a caplike "insulating plate" at the bottom of the bulb. This plate meets a springy metal strip in the socket called the "foot contact." Current leaves the bulb through this foot contact, which is connected to the second screw terminal and wire.

Sockets generally employ the same kind of screw terminals as plugs. Loop the stripped wire ends around the screw terminals . . .

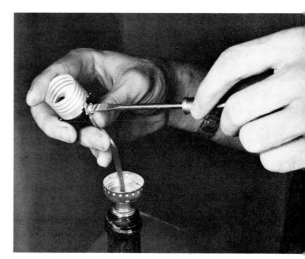

. . . and tighten the screws with a screwdriv-

Fit the wired socket mechanism into its housing.

The entire socket consists of A) an outer metal sheath, B) a cardboard insulating sheath, C) the socket mechanism with two screw terminals D), all of which sit in E) the socket's base. Often this base is threaded to screw onto lamp pipe F).

When connecting a row of lights along a single cord, wire them "in series" as at top, never "in parallel" as at bottom.

One convenient way of wiring a string of bulbs in series is with clip-on sockets. These sockets have tiny metal teeth that bite through the insulation, effecting the connection. Dozens of such sockets may be attached to a single line of wire. No stripping of insulation is required. However, the paired lamp wires must be split along the midline of insulation.

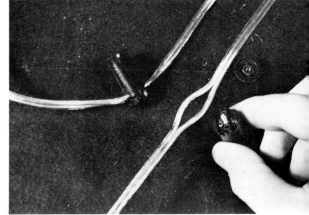

Series of Sockets

A series of sockets can be continued on a single cord. Strings of bulbs offer effective solutions in some hanging lamps, vanity lights, and Christmas tree ornaments. The critical point to be considered when combining a number of sockets on a single line relates to the wiring configuration. The sockets must be connected "in series," not "in parallel." "In series" means that for each socket on the line there is a pair of wires leading from the main pair of wires to the socket. (See diagram.) Many beginners try to string lights by connecting the first socket directly to the main cord, running some more wire from the screw terminals of the first socket to the next socket, and so on. This is known as "parallel" wiring. "In series" wiring allows the maximum amount of current to flow to each socket in the row, but "parallel" wiring divides the total current among all the sockets. Parallel wiring cuts down on the electricity flowing to each bulb, and dim, ineffective lighting results. There are sockets on the market that can be clipped directly onto a cord to produce a row of lights in series. Dozens can be put on a single cord. Small pin contacts in the base of the clip-on socket pierce the insulation of the wire, forming the small, in-series circuit.

To connect a simple in-line on/off switch, separate two inches of wire along the insulation midline. Cut one wire in the middle and strip about 1/2 inch of insulation from each cut end. Attach the two ends to the two screw terminals inside the switch housing as shown. Do not cut or strip the second wire.

In-Line Switches

The most common type of switch which can be put on a cord contains two screw terminals and an on-off button that opens and closes the circuit between these terminals. Before connecting a switch, be certain that the cord is disconnected from the wall power source. Remove the screws and nuts that hold the switch case halves together. At the point in the cord where the switch is desired, slit two inches of insulation along the line between the molded wires. Cut one of the wires and remove approximately 1/2 inch of insulation from the cut ends. Do not cut or remove any insulation from the second wire. Twist stranded wires into tight bundles. Form loops at the ends. Put the uncut wire into the slot provided for it in the switch housing. At the same time, wrap the wire loops around the screw terminals and tighten the screws. If it is difficult to get the loops to fit snugly against the screws, use pliers to squeeze the wire to the screw before tightening. Put the switch case halves back together, plug in the cord, and test the switch.

A second type of in-line switch—the thumb-wheel type—uses pin contacts instead of screw terminals. Slit the wire along the midline for about 3/4 of an inch in the area where the switch is to be installed. Cut one wire of the lamp cord near the beginning of the slit. *Do not remove any insulation.* Just nest the cord in the hollow half of the switch hous-

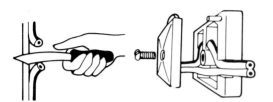
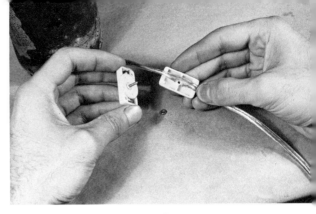

To connect a clip-on, thumb-wheel, in-line switch, again separate the cord into two insulated wires. Cut one of the wires, but do *not* remove any insulation. Just fit the wire into the housing as shown, and squeeze the housing shut. *Diagram Courtesy: LAMPARTS, Angelo Brothers Co.*

ing as shown. Fit the two halves of the switch together and squeeze so that the pin contacts pierce the insulation. Tighten the screw and nut which hold the casing halves together.

This same procedure is employed in installing a thumb-wheel dimmer switch. This dimmer switch has two positions: high and low. Its casing is virtually the same size as that of the pin contact on/off switch, but it provides more control over the flow of current and over the amount of light produced. A second type of dimmer switch which can be attached in line provides a full range of brightness control. Instead of just high and low, this switch can be adjusted for anywhere from 0 to 100 percent brightness. It is installed using the screw terminal method described above for the common on/off in-line switch.

Other Types of Switches

The other switches discussed earlier in this chapter—rotary, push button, pull chain, and toggle—are installed in fundamentally the same way. Instead of connecting the cord to screw terminals, however, the cord must be spliced and soldered to the wire leads coming out of the switch. The diagrams illustrate the wiring configurations and the uses of different switches. Each of the switches comes with two, three, or four color-coded wire leads, depending upon how many on/off combinations of lights are to be controlled. Leads from the switch should always be spliced, soldered, and taped to the lamp wire as discussed above.

From these switches, the black lead usually connects to one of the two wires extending from the plug. Through this lead, electricity enters the switch. The other leads are connected, in turn, to lamp wire which is attached to screw terminals on the sockets. The circuits vary according to the type of switch and the type of fixture. To complete the circuit, electricity flows through the switch, into lead wire(s) from the switch, into the bulbs, through the bulb filaments (producing light), and out the neutral wire leading back to the wall socket.

CONSTRUCTING AND WIRING THE BASIC LAMP   55

The simplest switch is the single pole, on/off type. Here it is shown hooked up to the two screw terminals of a light socket. One twist of the rotary switch and the light goes on; another twist and it goes off. The two leads coming from the switch may be reversed. This identical wiring is applicable for push-button, pull-chain, and toggle switches as well as for rotary switches.

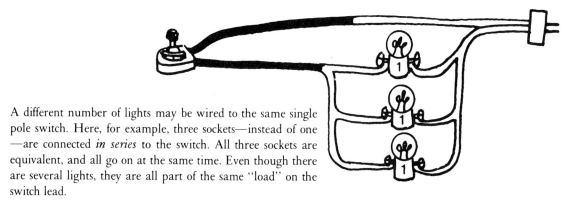

A different number of lights may be wired to the same single pole switch. Here, for example, three sockets—instead of one—are connected *in series* to the switch. All three sockets are equivalent, and all go on at the same time. Even though there are several lights, they are all part of the same "load" on the switch lead.

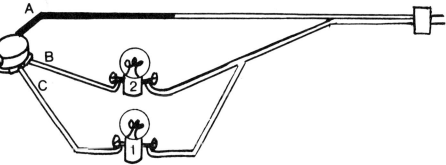

A more sophisticated switch allows two loads to go on and off independent of one another. Called a two-speed pull-chain switch, this mechanism has a black lead A) connected directly to the power source, a blue lead B), and a red lead C). The red and blue leads are in control of two separate loads. [Note: These lead colors are based on Leviton brand switches and may vary according to the manufacturer.] With one pull of the chain, load 1 goes on. With a second pull, it goes off. With a third pull, load 2 goes on, and with a fourth pull, it goes off. Note that while only one lamp is shown in loads 1 and 2, actually any number of sockets could be linked in series to either or both loads. Two-speed switches are available in switch types other than just pull chains.

The three-speed, four-position rotary switch handles three separate loads with four color-coded switch leads. A), the black lead, connects to the power source. B) is red, C) yellow, and D) blue. With one twist of the knob, load 1 goes on. With two twists, load 2 also goes on. With a third twist, load 3 joins 1 and 2 in the on position. A fourth twist turns all loads off.

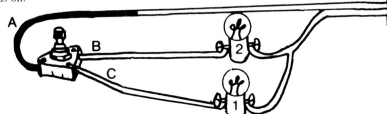

With the two-circuit, four-position rotary switch, loads 1 and 2 can be turned on separately or together. A version of this switch is often used for pole lamps. The A) black lead is connected to the power source; B) is blue and C) is red. With one twist of the switch, load 1 goes on. With a second twist, load 1 goes off while load 2 goes on. With a third twist, loads 1 and 2 go on together. The fourth position shuts off all lights. The switches shown in the preceding diagrams represent just a few of the many types of switches available. But with these simply wired switches virtually any lamp job can be satisfactorily completed.

Ceiling and Wall Hookups

The lampmaker is restricted in the installation of wall and ceiling fixtures to those room locations where junction boxes already exist. It is *not* advisable for the novice to attempt to install electrical boxes in walls or ceilings. However, it is easy to construct wall and ceiling lamps which can replace fixtures currently installed at junction boxes.

Always be certain that the main circuit breakers or fuses are off before working on the circuit. To take down the old fixture, first remove the shade, bulbs, and other lamp accessories. Undo the nut(s) which hold the ceiling plate or canopy to the wall or ceiling. Under this cover plate is a hole, and coming from the hole are white and black (or colored) wires. The actual number of wire leads is dependent on the number of wall switches which control the fixture. In disassembling the old lamp, keep track of which wires were connected where. The easiest way to be certain that you have properly wired a new ceiling fixture is to pay strict attention when unwiring the old lamp. Virtually without exception, same color wires go together: white to white, black to black. They are typically twisted together in a pigtail-style splice. When connecting wires from the new lamp to the ceiling box leads, attach one wire from each socket in the fixture to the black leads, and the other

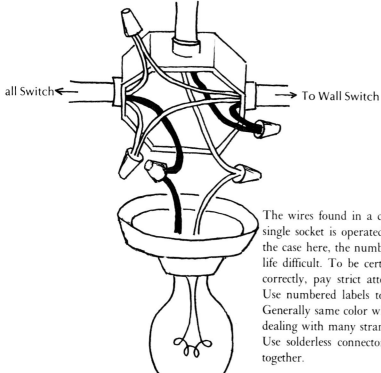

The wires found in a ceiling box can be confusing. When a single socket is operated by two separate wall switches, as is the case here, the number of wires to be connected can make life difficult. To be certain that you are wiring a new lamp correctly, pay strict attention when unwiring the old lamp. Use numbered labels to mark which wires belong together. Generally same color wires are combined—but when you are dealing with many strands, this rule does not necessarily hold. Use solderless connectors to hold the exposed ends of wires together.

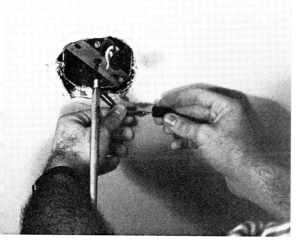

Solderless connectors are screwed onto pigtail-spliced wires.

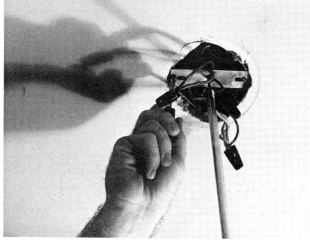

The 4-inch crossbar is screwed to lugs in the ceiling junction box.

A threaded pipe, or steel nipple, may be attached through the center of the crossbar with an hexagonal nut.

wire from each socket to the white leads. Use the cone-shaped wire nuts (solderless connectors) to effect the connection.

Screw or bolt the lamp fixture to the ceiling. Install bulbs. Turn the main circuit back on. If the light doesn't work, first make certain that the bulbs are good. Then turn the main power off and disassemble the fixture. Check all connections to see 1) if the wires are crossed (white goes to white, black goes to black), 2) if all the wires are in contact with one another (remove each wire nut, give the splice a twist, and replace the wire nut), and 3) if all points of connection are properly insulated (be certain that no stray wires or metal from one terminal come in contact with other terminals).

Fluorescent Fixtures

There are primarily four parts to be installed in a fluorescent lighting fixture. When purchased, the fluorescent fixture generally comes with a starter, a pair of end sockets, and a unit called the ballast. Tubes are purchased separately. All wiring is performed using wire nuts or screw terminals.

To install the starter, put it in its slot, and turn it 1/4 revolution clockwise. Mount the ballast to the back of the lighting box, using the screws and nuts supplied. Screw the two end slots in place as shown. Perform the indicated wiring of all leads to and from the end slots and ballast. There are several different "knockout" holes through which wires may leave the box. The fluorescent unit is readily attached to wall or ceiling box leads, or it can be spliced into regular lamp cord and plugged in.

Before screwing the cover plate on the fluorescent lighting box, install the tube. Check to see that the fluorescent tube goes on. If it works, remove the tube, screw down the cover plate, and reinstall the tube. If the tube will not light, check all connections. If the unit is an old fluorescent fixture you are renovating, it may need a new starter, new ballast, or new tube. Check the tube first, the starter second, the ballast last. If the tube blinks on and off, check to see that the tube is properly fitted in its sockets. It may be necessary to gently sand the contacts on the tube to improve contact. If blinking persists, check the starter and ballast. If the blinking seems to swirl around in the tube, and if this persists over a long period of use, replace the starter. There are several discoloration symptoms of fluorescent tubes. If a new tube is slightly brownish, that is considered normal. But if it turns a darker color at only one end, reverse the tube. If the tube discolors on only one side, turn the tube over. To eliminate any buzzing sounds, check to see that connections to the ballast are all tight. If the noise continues, replace the ballast with a "low noise" ballast unit.

CONSTRUCTING AND WIRING THE BASIC LAMP 59

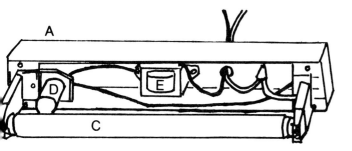

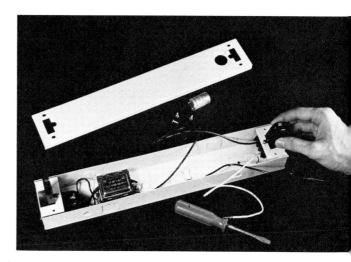

The fluorescent fixture comes in a white metal box A). Two end slots must be screwed in place B). They hold the fluorescent tube C). A cylindrical starter D) is installed in its designated slot by giving it a quarter turn. The ballast E) is bolted in place at the back of the box. Wires are connected by screw terminals or solderless connectors. Wires may leave the box through punch-out holes in the back or sides.

3

Lampshade Making

Anyone who has had any experience in shopping for lampshades will have discovered that shades are very expensive and that sometimes the most suitable shade is difficult to find. Yet making, or renewing, lampshades is not very difficult, is very inexpensive, and professional results can be achieved with a bit of experience.

The lampshade is a lighting essential, particularly for table and floor lamps. It is a light controller—concealing, diffusing, and reflecting light, as well as an element in interior decor.

KINDS OF LAMPSHADES

Lampshades vary in size and shape, structural design, and kinds of material used for coverings. Different combinations will suit different styles of lamp bases and will coordinate with different periods of furnishings. There are few traditions in lampshade designing. Most of them are comparatively recent in origin. One is the chimney shade, originally constructed to fit over kerosene and gas lamps. Another is the Tiffany style lampshade, and still another is the chandelier converted from candle or gas flame to electricity.

Many lamps are shades that function as a complete unit. (Many of these designs are shown throughout the book.) Usually they are table, floor, or ceiling lamps.

Then, there are soft and hard lampshades. The soft shades are constructed of fabric that is stretched and sewn onto the frame; hard shades are made of semirigid materials such as plastics, parchment, metal, vellum, and fabrics laminated to plastic backings. Rigid materials do not require a frame with struts as fabric does, just top and bottom rings.

Some Lampshade Styles

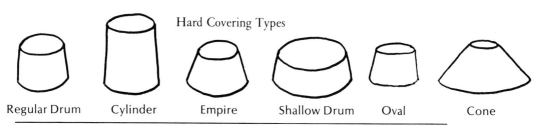

Hard Covering Types

Regular Drum · Cylinder · Empire · Shallow Drum · Oval · Cone

Anatomy of a Shade

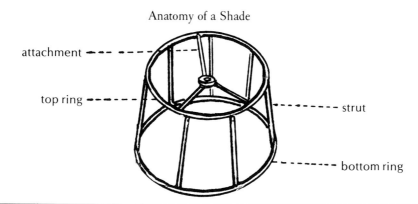

attachment · top ring · strut · bottom ring

Some Lampshade Attachments

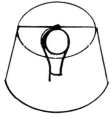
Butterfly Clip

Uno-Bridge

Washer

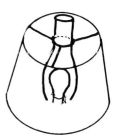
Chimney

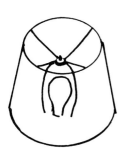
Partially Recessed Washer

Recessed Washer

Gimbal Fitting (England) allows shade to be tilted

Ring Louver

Devices for supporting shades are constructed to accommodate a variety of different lamp designs. Some shade fittings snap directly onto the bulb by means of a butterfly clip. Other shades are equipped with washers that fit over the male members of a harp. A finial then screws over the washer, securing and punctuating the lamp. Uno-bridge attachments screw onto a socket or slide over the glass chimney of a lamp. Each of these attachments can be flush with the top ring of the frame or recessed (into the shade) in varying depths. In England, gimbals, which allow the shade to be tilted, are available so that one can focus the light for special tasks—like reading in bed.

Contours of hard-sided shades are limited to whatever can be done with a somewhat stiff material. Sides usually are flat, pleated, scalloped, or scored-cut planes, much as in making a paper sculpture. Frames for fabric coverings are available in an endless variety of sizes and shapes—petaled, scalloped, in-and-out bell, pagoda, V-notched—and combinations of these shapes, along with standard types used for hard sides (or soft): regular drum, shallow drum, cylinder, oval, hexagon, cone, square, rectangle.

Combining these elements in different ways allows a wide range of choice. Where there is no design solution using standardized frames, lampshade frame makers will construct shade frames to order. Some lampshade shapes have been standardized over the years according to stylistic influences.

DESIGN ASPECTS

Proportion

Proportion or scale is the most critical aspect in designing a lamp. The size, shape, and style of a lampshade can transport a lamp to the sublime or the ridiculous. People seem to have the most difficulty assessing proportion when selecting shades. To be on the safe side, most people duplicate shade replacements and don't consider other possibilities that might have been better solutions—particularly since we have new attractive shade coverings.

Perhaps the reason for so much difficulty in making aesthetic judgments is that there are no firm guidelines for estimating the relationship between shade and lamp base. Anything that looks good (to the experienced eye) works well. If any rule of thumb is helpful, it is that a shade usually looks best if it is about as tall as the base of a table lamp, give or take an inch. If a base is 16 inches tall, then its shade should be about 16 inches tall. But there are many exceptions. If a lamp base is very

Antique bronze table lamp with fabric on vinyl hard shade. Note that of the 32-inch height, half is the length of the shade.
Courtesy: Keystone Lamps

An oak base and white linen hard shade, 24" in diameter. The total height is 36"; half is the height of the shade.
Courtesy: George Kovacs Lighting, Inc.

tall, a similarly tall shade could build the combination of lamp and shade to monumental heights. Instead, a workable solution is to make the shade wider and more shallow.

Shades that are too small look like top hats, and, conversely, shades that are too big overwhelm the lamp base and the lamp looks like "all shade."

Suitability of Coverings

Another aspect in design selection is suitability of shade shape and covering. Fabric-covered shades are more traditional in feeling and look well on traditional bases, particularly glass, pottery, porcelain, and metal materials, as well as over novelty forms such as figurines, ginger jars, and decorative cylinders (like converted embossed metal rollers formerly designed to print fabrics).

Burlap and rough nubby materials would not be smooth enough to be compatible with finely detailed porcelains and the like; whereas burlap, coarse linen, and new textured plastics are very effectively combined with bases of wood, pottery, some plastics, metal, and so on.

Scale

Scale is an element that can offer surprises because we recognize scale only in relationship to some other item. It is, therefore, very difficult to determine whether a lamp will be sized in proportion to suit furniture scale and surrounding space unless a ruler or some comparably sized object is used as an aid to prejudge or premeasure. People who have lived in small spaces and then move to larger quarters often discover that their old furnishings look puny and completely out of scale. If there are no reference points around for you to make relative size judgments in designing a lamp or its shade, then get a ruler and determine size parameters.

Function

The function of a lamp can limit what its shade can look like. A lampshade for a reflector bulb will have to be of different construction and material to accommodate for the heat of the bulb and the need to focus its light. A lamp containing multiple bulbs will necessarily sport a wider diameter shade than a single bulb lamp. And a "shade" for a fluorescent lamp would be very different in design from one for a boudoir lamp. (See the Hand Lamp in chapter 5.)

Of course, bulb safety plays a "functioning" role in lampshade design. Lamps should be rated by the safety limits in wattage. Many commercial lamps are so categorized. But how can we determine bulb size when we design a lamp ourselves? One way is to test the proposed bulb next to the desired lampshade covering. Allow the proper distance between material and bulb, as well as a reasonable amount of time. If bulb heat affects the lampshade material by turning it yellow, brown, melting or distorting, creating cracks, and so on, then reduce the wattage, change bulb shape and/or size, and consider changing the proposed shape of the shade. Besides changing bulb wattage and size, another solution is to allow for venting of bulb heat; still another is to increase the distance between bulb and shade material.

Style

Style is intrinsically related to the decor of a room. Some people prefer ornately decorated bases and shades, thinking that it provides an opulent accent. We feel that simple, classical forms blend better with any period. Our preference is very definitely toward simple contours and contemporary design.

There are some classics, however, that blend well in almost any setting—one is the ginger jar with its pleated shade; another is a simple pottery vase shape with a cone or drum shade. The hourglass or hurricane shape as lamp base or undershade form is also a classic. So are some of the Tiffany style lamps or shades.

Innovation

Innovation for the sake of creating a unique design—just to innovate—rarely works. A classic-shaped shade probably would function better. But where a solution to shielding light is elusive, new designs certainly are in order.

When the plastic shade on a column lamp melted, it signaled the need for a new lampshade solution. Some ventilated, nonmeltable material would provide the best answer. A "found" frame was discovered in a gardening supply store. It actually was a dome fashioned out of enameled wire (used to hang sphagnum moss as a hanging planter or used as an ivy training frame), which would rest on the reflector of the lamp without need for a shade adaptor or harp. The size and shape suited the lamp admirably.

While shopping in building-supply supermarkets, hardware stores, or plastic specialty shops, many products and materials designed for other purposes will strike you as adaptable. And old lamps can be renewed with new shades—different in covering, size, and shape. Basic shades can also be personalized by decorating them by using paint, fabric, paper, or ribbon. (More about this later.)

Frames can also be made by forming wet reed (as used in basketmaking) into petallike units, gluing them together, and then applying fabric or raffia cloth to the petals.

GENERAL ASPECTS OF MAKING A BASIC LAMPSHADE

After assessing what shape and size of frame are necessary, the material selection follows. Here the range is tremendous. Choice ranges from soft to rigid materials. Of the two types, soft covered shades are the more difficult to construct.

Hard-backed shade coverings come in a wide variety of textures, patterns, and colors. These can be attached by tape backed with a water-soluble glue.

In some hard-backed shade coverings, textured fabrics are laminated to vinyl plastic backings as in the examples shown here, except for vinyl that is metalized and embossed as in the sample in the lower right and the embossed example, right center. These vinyls are usually applied to shade rings with a chemically activated tape and use of mechanical pressure.

Material Selection

Hard-shade covering materials may or may not be specifically made for lamp covering. There are many semirigid art materials, particularly those discovered in plastics supply stores, that are eminently suitable. These may be vellums, stiff decorative papers, acetates used in color separation work, and so on. Some fabrics, papers, and plastics are clad or laminated under heat and pressure to a vellum or plastic backing and sold as shade covering. The selection is broad, varying from linen and burlap laminated over or between acetate, embossed plastics, vinyl, fiberglass laminations, various fabric-embedded materials; linen, straw, slats, and bamboo are some other textures that are available. Most of these sheet materials are richly textured and complementary to lamp base structure and design. It is possible to use fabrics (matching draperies or curtains) and glue them to a vellum backing. Use a water-diluted solution of white glue such as Elmer's® or Sobo®, a similar adhesive such as Velverette® (used here) or an iron-on webbing like Poly-Web®.

If a shade material is made completely of plastic and does not have a paper or vellum backing, then special techniques must be used to mount the plastic-backed material to the metal frame of the shade-to-be. Commercially, this is done with chemical tape—a tape that requires activation by a solvent and attachment by a machine that applies pressure. Another solution is to draw a narrow bead of powerful adhesive, such as a clear two-part epoxy (there is a type that hardens in five minutes), along the edge and apply moderate pressure until the adhesive hardens; or punch holes along the edge and attach the plastic onto the frame with a decorative lacing stitch.

Soft materials range from light to medium-weight fabrics of just about any substance—linen, silk, acrylic, Celanese, nylon, and so on, in any pattern or color. Most popular among these are smooth silks, shantungs, taffetas, eyelet, chiffon, crepe; most are in pastel colors such as white, eggshell, ecru, honey, gold, light green, and so on. Although these are most popular, there is no limit to what could be suitable.

Not all, but most soft-covered shades are lined, usually with a finely textured, lightweight rayon or silk lining material in a neutral color such as white or ecru.

The amount of fabric necessary can be determined by measuring the pattern.

Patternmaking for Drums, Cylinders, and Cones

Hard coverings have to be accurately cut to the exact size in order for the material to fit around the two frame rings (one for the top and one for the bottom). The ring at top is used for mounting the shade to the base as well as giving an unchangeable form to the lamp; the bottom ring is necessary, in most shades, to maintain the lampshade shape at the bottom of the shade.

There is a useful, easily executed formula for creating patterns for shades in which one end is smaller than the other. This is the simple equation:

$$\text{Pattern Top Radius} = \frac{\text{Side Length} \times \text{Top Ring Diameter}}{\text{Bottom Ring Diameter} - \text{Top Ring Diameter}}$$

Let us say that the length of the shade is 12 inches, the diameter of the top ring is 6 inches, and the bottom is 12 inches. Then the radius of pattern top is 12 inches times 6 inches divided by 12 inches minus 6 inches or:

$$\text{Pattern Top Radius} = \frac{12'' \times 6''}{12'' - 6''} = \frac{72''}{6''} = 12''$$

With a 12-inch radius, we can trace off the top arc of the shade pattern by cutting a string 12 inches long, taking one end and, with a pencil attached to the string (make allowance for cord around pencil), drawing an arc on a large piece of paper. This is a makeshift compass. This provides the top arc of the pattern. To form the bottom perimeter of the shade pattern, add to the top radius the length of the shade.

Pattern Bottom Radius = Radius of Top + Length

or

R = 12'' + 12'' = 24''

Then, maintaining the same tack point as for the top arc, draw a second arc with a pencil tied to a 24-inch length of string.

To determine the length of the pattern, indicate a starting point on the top ring and on the pattern with a marker or piece of tape and roll the ring along the arc line. When the starting point rolls around, allow

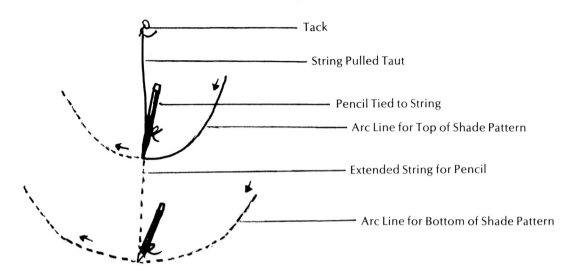

one or two inches of overlap—trim away the excess later. Repeat the same operation for the bottom arc line using the bottom ring. With a ruler, draw a height line from the top arc to the bottom arc to indicate shade length. For the measurements indicated in our example, it should be 12 inches from the top to bottom at each end. If you have gone over or under, adjust the "compass" string to make correction.

Patternmaking for Square, Rectangular, and Hexagonal Shades

Rectangular, hexagonal, and square shades are usually constructed with struts and can be made by tracing pattern dimensions directly on paper or by using a ruler to transfer measurements from frame to pattern paper.

If the sides or corners are not exactly square, but each side is the same —because the top "ring" is smaller than the bottom "ring"—a continuous pattern might be desirable.

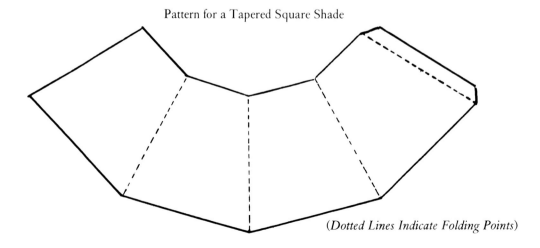

Pattern for a Tapered Square Shade

(Dotted Lines Indicate Folding Points)

Trace off a single square on a piece of cardboard. Cut it out. Then, on your large sheet of patternmaking paper, trace around the square. Lift the pattern and juxtapose the edge to the line previously drawn. Trace a second side. Continue, until all four sides are drawn. You will note that they are not strung out but, rather, are tangential to an arc.

Creases can be scored for precise folding. Make certain that some overlap is allowed on the last piece so that the open edges or seam can be glued together.

Constructing a pattern for soft-covered shades is not necessary, as you will see later when the process is described.

Preparation of the Frame

Almost all frames are metal and require any sharp edges or joints to be painted or sprayed and/or bound or wrapped before coverings are applied.

If there are any sharp edges where the rings or struts are attached, remove them with a fine file, filing in one direction—up and away from the joint.

Some metals tend to rust, so a spraying with clear Krylon® (a clear acrylic spray) or painting with a fast-drying paint the same color as the fabric is a good idea. Some frames come prepared for covering.

Struts may or may not need to be bound. Attachments at top of shade are never bound. A protective coating may be enough. Traditionally, struts were covered with a bias binding in a neutral color or a bias binding cut from the lining material. (To cut a bias, find the diagonal of the fabric by pulling it until it stretches and then cut along the bias path—not along straight edges of the warp or weft.)

It is essential to cover the rings of soft-covered shades, however, because the covering material is sewn to the very tightly wrapped binding. The binding should be wound so tightly that it will not twist or slip even a fraction of an inch.

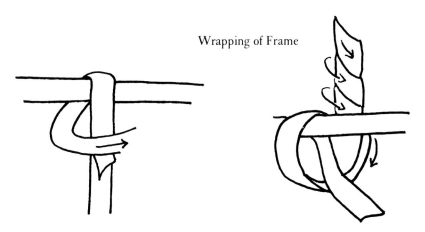

Wrapping of Frame

To begin binding a strut   To end binding a strut, form a knot

Rings of hard-covered shades may be covered or not. It is just as well to glue the hard covering to the rings with a binding of some kind.

Binding tapes may be of various materials—glue-backed paper, ribbon, bias binding, strips of self-material (the material itself cut into binding). Use 3/8-inch tape for rings, 3/8-inch or 1/4-inch for struts. If struts are to be bound, bind them before the rings.

To bind rings or struts, wrap the frame tightly and overlap the ends slightly. As with struts, do not allow the fabric to become twisted and keep the binding taut while wrapping the frame. One way to achieve this is to hold the tape at an acute angle to the wire. You will probably require a length of tape that measures twice the circumference of the ring (or strut as the case may be). When the binding has been completely wound, either glue or stitch the end of the binding tape to the beginning part of the tape. At all times keep hands clean and free of grease.

Another method of binding is to wrap the tape along the frame by following the contour and not winding the tape around the wire.

Paper tape that needs to be moistened with water can be used in the winding method or by wrapping it along the frame. To moisten the tape, run it over a wet sponge. Then press the tape firmly to the frame with fingers so there are no gaps.

Applying Hard Materials

Some hard materials are translucent and permit light to transmit through the shade in varying degrees, others are opaque. Parchment and buckram were early hard materials, but more recently there is a wide range of plastics—acetates, vinyls, fiberglass, polyester—that have become available. Many are commercially bonded.

In making a pattern for an oval frame (completed shade shown here), mark a starting line, as seen on the shade, and rotate shade along pattern paper while tracing the route made by the rotating lampshade.

Repeat the same procedure for the bottom arc. Cut out the paper pattern and try it out by clipping it to the top and bottom shade rings.

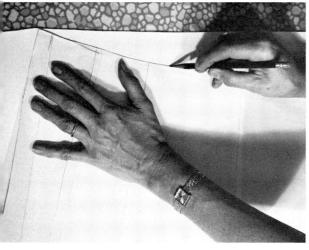

When satisfied that the pattern fits, trace around the pattern on the back of the shade material. Then cut out the pattern.

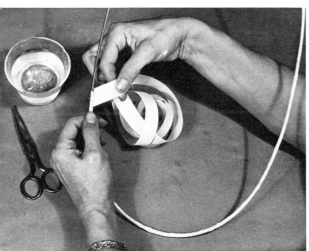

Tightly wrap water-soluble-glue-backed tape around both the top and bottom shade rings, overlapping each previous wrap very slightly. (Beginners tend to wind too thickly.) Binding should be tight and smooth. If, with a thumb and forefinger twisting test, the tape twists in either direction, it should be done over again.

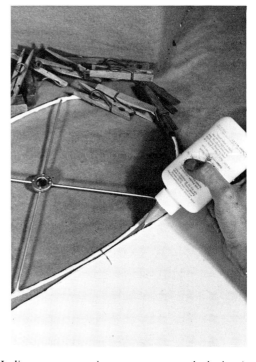

Indicate center point on pattern and shade ring. Draw a bead of white glue on the shade covering back. (This ring, by the way, is a four arm washer fitting.)

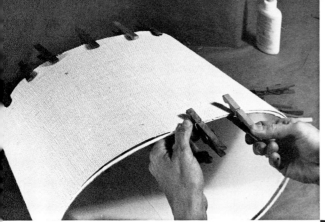

Attach both rings to the shade using spring type clothespins.

Extend a bead of glue at the seam where the shade material overlaps.

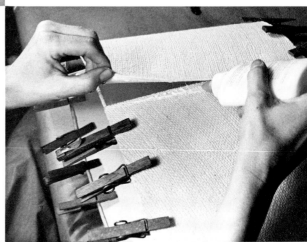

Weight the seam down with weights, strips of wood, or whatever will keep the seam flat and under moderate pressure on a flat surface.

A bead of white glue is run along the top and bottom edges in preparation for attachment of trim.

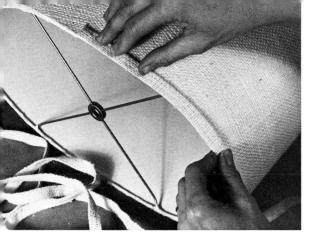

Trim of the same material is attached. Note that the trim begins and ends at the same seam where the shade covering overlaps. This shade can be seen on the wooden sculpture lamp in the beginning of chapter 4.

Most wallpapers and fabrics can be glued to a vellum, parchment, or bristol board background with Velverette®, a white glue that has the proper viscosity for adhering fabrics, or with fusible polymer webbings such as Poly-Web®. Papers and fabric can be ironed onto a semirigid backing. The heat of an iron melts the webbing, assuring an even sticking. The only fabrics that would be difficult to iron onto a backing would be high pile materials. These become shiny when pressed. In England, one can purchase an adhesive-backed cardboard called Parbond. When wallpaper or fabric is placed over the adhesive and pressed with an iron, the heat of the iron melts the adhesive on the cardboard, thereby adhering the fabric. Wallpaper, decorative paper, and fabrics can also be attached in decorative patterns.

After selecting the covering and tracing around the pattern (on the wrong side), cut out the shape along the penciled outline.

To assemble the hard material to the frame rings, start with the bottom one first. Spring-type clothespins are used, universally, to clip the material to the frame. The ring should not be visible when looking at the side of the shade, nor should it be up too high. Rather, the bottom ring should be level with the bottom edge.

It is possible to sew coverings to the ring, using an overstitch or whipstitch, but use of an adhesive such as Elmer's Glue®, Sobo®, Uhu®, Evastick®, and so on, will yield good results on a paper-backed covering. A more powerful adhesive, such as the two-part epoxy mentioned earlier, would be necessary for plastic or plastic-backed coverings.

Apply the adhesive sparingly to the outside of the ring. Using spring-type clothespins, attach covering to ring. The stiffer the covering, the closer together the pins must be. Do not remove the clothespins until the adhesive is quite dry. Attach the top ring in the same manner.

Next, carefully insert a small amount of adhesive between the seam overlap (overlap about 1/2 inch to 3/4 inch). An applicator stick or large darning needle can be used to apply the glue. Then place a piece of polyethylene on the outside and inside and lay the shade seam-side down on a table. Place a weight over the seam so that a flat joining will result. The polyethylene can be pulled away later (even if there is oozing)—most glues won't stick to it.

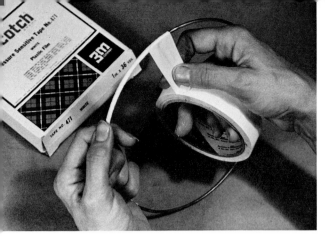 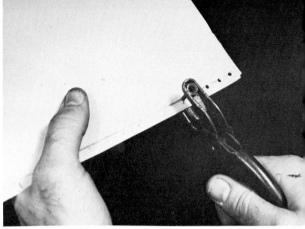

This shade ring is wrapped with a pressure-sensitive white plastic tape in preparation for a white acetate cover that is to be laced onto the ring.

With a hole punch, holes are punched along the edges at evenly measured and marked points.

Using plastic lacing, the stitch is whipped along the edge, thereby tying the ring to the covering.

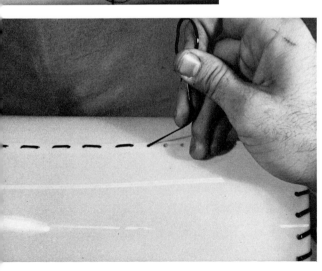

The completed hanging lampshade.

Then, with a running stitch, the lacing is threaded in and out and pulled tightly to sew the edges together.

Alternative ways of lacing provide decorative edging effects.

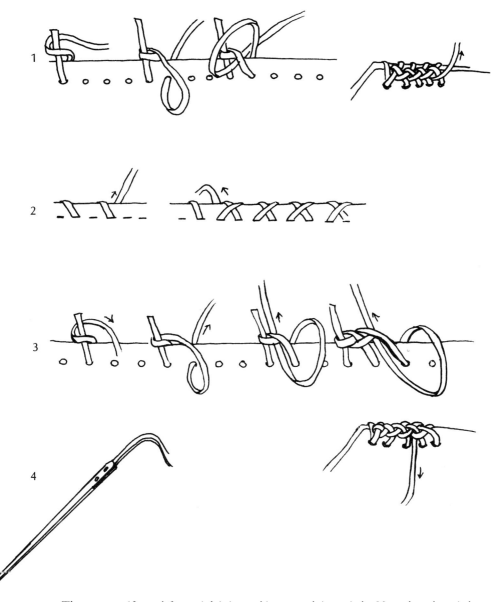

1. Three steps (*from left to right*) in making an edging stitch. Note that the stitch moves principally in one forward direction.
2. To create a simple cross-stitch, first use every other space in one direction and then reverse the direction, filling in the skipped spaces while crossing the other stitches.
3. Four steps in making a double buttonhole (or double layover) lacing stitch. Note that the lacing returns on itself.
4. A lacing needle (purchased along with lacing or thonging where leather supplies are sold). Note that lacing fits into its two-pronged slot. Use of a needle is optional. The plastic lacing shown above does not require a needle because it is stiff enough to be threaded without one.

The last operation in making a shade is applying trimmings. But before we get to that, let's look into covering shades with soft materials.

Applying Soft Coverings

The most applicable fabrics for soft shades are those that can be stretched somewhat (have "give"), are strong without being bulky, are not easily marked with pins, do not fray, and for most instances are translucent. Silks, linens, shantung, some cottons, thinner drapery fabrics, nylon, to name a few, work very well. This excludes a wide range of excellent materials too numerous to mention.

PINNING

Generally, it is not necessary to cut exact patterns for soft coverings, just approximate sizes. In making soft fabric shades the fabric is pinned to the bound frame in order to fit either the covering or the lining, preparatory to stitching. Pins are placed horizontally through the material and binding of the frame.

STITCHING

To attach covers or linings to a frame after pinning, use either an overcast or a glovemaker's stitch (shaped like a Z). One hand pulls the material taut while the other inserts the needle and thread. A quilter's thread and standard needle work well. Thread lengths should be short to avoid tangles, and stitches should be about 1/4 inch apart. The width of stitch should be as narrow as possible, but this is determined by the width of the final trim which is sewed or glued over the stitching later. Only a small amount of material should be picked up as the needle passes through material and binding (of the frame).

If the fabric is tough, then use a running stitch. The Z-shaped or glovemaker's stitch is made by piercing the needle through fabric and binding and then reinserting it in the same hole for the beginning of the next stitch.

If a fabric tends to ravel, then paint a bead of white glue along the edge and wait for it to dry before attempting to attach the fabric.

Each section can be sewn this way, attaching the fabric section by section to the struts. Take care not to create bulk at the seams by overlapping too much, or else a very wide trim will be necessary later to mask the sewing.

When each section has been stitched, trim excess fabric away. If there are signs of raveling now, paint the edges with slightly water-diluted white glue.

Bias tape is wrapped around an ivy-training frame only on vertical struts. (Source is garden supply center.)

An open weave linen fabric is pinned to each segment of the frame. The pin pierces the tape.

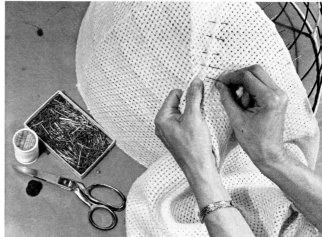

Using a whipstitch and short lengths of strong quilter's cotton, the fabric is sewn to the tape for each section. As the stitching is made, the pins are removed.

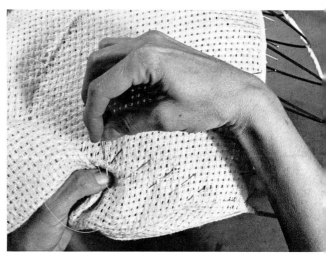

Excess fabric is trimmed away.

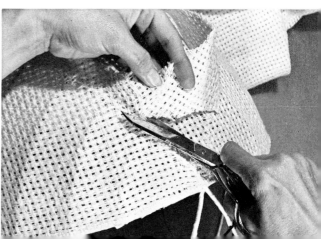

A bead of white glue is run along the seam in preparation for adding the tape trim. The glue also seals the fabric edges and fixes the stitching.

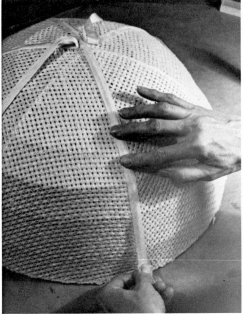

The tape is pressed along the glue and pulled taut at the same time.

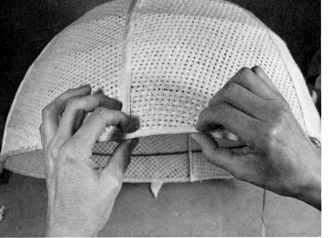

Along the edge, the tape is folded around the bottom ring of the shade.

This shade replaced a plastic shade that had melted because of the excessive heat of the bulb and low melting point of the plastic. Note that the horizontal rings of the shade create an interesting pattern when the lamp is lighted.

Tailored Soft-Covered Shades

In case your decision is not to sew the covering to the struts but to stitch seams by machine (or by hand), proceed as usual through the pinning stage and then with tailor's chalk mark through the pins, indicating sewing lines. Then remove the pins and cut the fabric about 1/2 inch from the tailor's chalk lines all around. After all gores or parts of the fabric have been cut, using a fine stitch, by hand or machine, sew along the seam. Seam widths should be even and pressed either to one side or open because they can be seen when the lamp is turned on. If no lining is to be used, then use a French seam, sewing the first part of the seam with the right side of the fabric out. Then clip the seam back to as narrow a width as possible and sew a second seam just outside the edge of the seam, this time on the wrong side of the fabric. No raw edges will show in this very strong seam.

After all parts have been sewn together, slide this tubelike cover over the frame, pin, stretch, and stitch the cover to the bottom ring, and then to the top ring. No puckers should be seen because the fabric should be pulled taut. Trim away excess fabric with scissors, and the shade is ready for trimming.

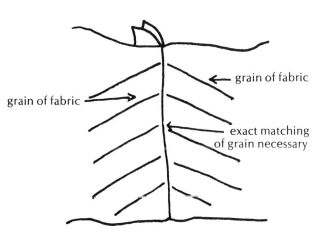

Grain Matching of Tailored Soft Shade Covering

A potichomania base by June Meier of the Cricket Cage and a soft tailored shade of silk shantung. (For potichomania technique see chapter 4.)

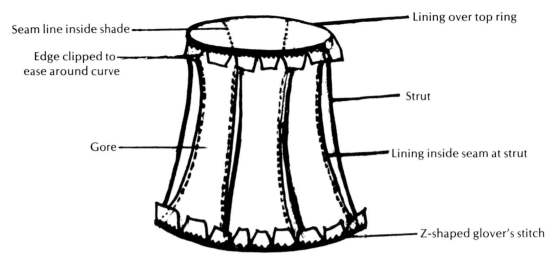

Lining of Tailored Soft-Covered Shade

Linings

Some soft-covered shades look better with linings. In that case, all the frame is hidden between layers of fabric.

Linings are made the same way as tailored soft-covered shades. Linings, however, are attached first, before the covers. The only change is that at the top and bottom rings all stitching appears on the front, to be hidden later by the outside covering and trim; therefore, at the top and bottom edge the lining must be a bit longer so it can be rolled around the top and bottom rings and be pinned. Then it is sewed to the binding at the outside edge.

Sometimes the lining has to be clipped at curves and along the top and bottom edges to ease stretching and prevent gathering and puckering.

When placing the cover on a lined frame, seams should overlap and be at the same point as the struts, so that when the lamplight is turned on there is not a confusion of seams. Also make certain that both lining and cover are tightly stretched so that there is no puckering or gathering.

Trimmings

Perhaps the two most important steps in making a lampshade are initially binding the frame (wherever necessary) and, secondly, applying the final trim. Trimming functions in two ways, it conceals stitches and raw edges and provides a decorative touch. In the case of hard-covered shades, trimming can be at once decorative and help to bind or attach the covering to the frame.

A small assortment of different kinds of trims that could be applied as shade trims. Trims vary from dress and coat trims to upholstery trims to various kinds of braids and ribbons.

Another method of attaching hard covering to rings is by using glue, tapes, and trims. Here the covering has been glued to the struts of a rectangular frame.

The edges are trimmed with a sharp knife.

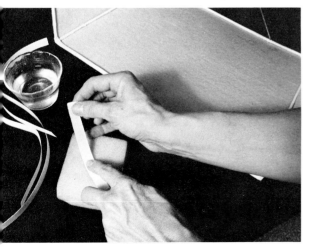

The glue backing of a paper lampshade tape is run over a damp sponge . . .

. . . and the tape is folded along the strut corners.

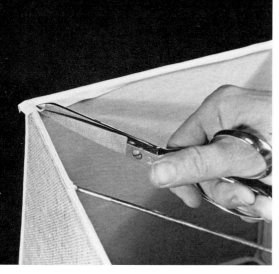

At the top and bottom edges, the tape is wrapped around the edge and to the top and bottom rings. This helps to attach and reinforce the cover. Here a corner is clipped to make for a neat corner.

After all edges are taped, white glue (Velverette) is applied over the paper tape.

Ribbon is folded around corners first and then around the top and bottom rings to cover the tape.

Corners are clipped to ease pulling.

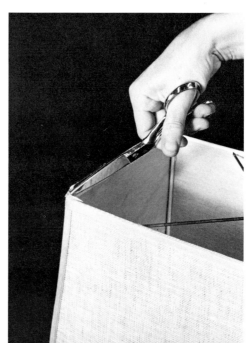

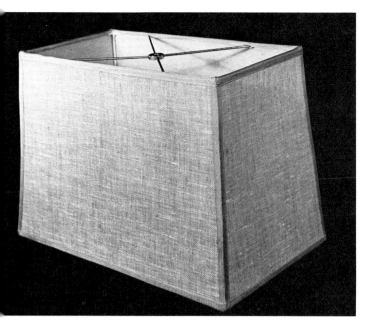

The completed shade.

The range of trimming available is large. There are chemically activated trims that are used commercially, paper, pressed, or fiber trims that are glue-backed and made especially for lampshades; and there are a wide variety of braids and ribbons, laces, fringes, and tapes as well as trims made of self materials. Selection should be governed by what looks best, of course, and the colors of the room. Perhaps the safest trims to use are those that match the covering. Where match is difficult, contrasting trim is advisable. If a shade is to be washed, make certain that the trim is washable as well. The width of trims should vary in proportion to the size of the shade. Three-eighths of an inch wide, however, is an acceptable width for most shades.

PLACEMENT OF TRIMS

Struts and rings are not attractive, so the trim should be placed where it will do the most for the shade. In order to mask the rings, some trims are rolled around the edge of the ring, others need to be positioned exactly at the edge. Trim on struts should be applied first—before the top and bottom edges are covered. When covering top and bottom rings, start the trim at a seam or strut and tack it with a pin.

ATTACHING TRIMS

Most trims can be glued in place. When the type of material will not permit gluing, invisible tacking with a needle and thread using a blind stitch (behind the tape, and between tape and shade covering) will do the job. Velverette® is perhaps the most satisfactory adhesive because it does not penetrate deeply through fabric or paper. But Elmer's or Sobo —any white glue—will perform fairly well, too.

To attach the trim, pin the edge in place at a seam and run a thin bead of glue along the line or edge to be trimmed. Then, with fingers, carefully press the trim into place, pulling it slightly to avoid bunching or scalloping. If the shade is unlined, clip clothespins along the trimmed edges to hold the trim securely while the glue dries. This is optional. The trim should, of its own weight, lie flat on the covering without moving if the shade is not disturbed. By the way, whenever handling a frame, hold it with the flat of your palms, one hand on the top ring and, if necessary, one on the bottom ring or by the attachment center of the top ring. Tension can be lost if the shade is held improperly. When edges meet, trim the excess away with scissors. If necessary, fold under the edges and glue them or stitch both ends together. The kind of material used and which method produces the neatest and longest-lasting job are the determinants of which technique to use.

VARIATIONS IN LAMPSHADES

It is possible to change or redesign a basic shade, or even renew one you already have, with various decorative techniques. It is also possible to vary the basic shade design with fluted or pleated shades, or by material —use of leather or copper.

Overlays

Wallpaper, fabric, colored acetate, ribbon, bows, and colored parchment can be cut into shape or designs and glued to a basic shade. Sprament® (a sprayable rubber cement), white glue, and colored tapes can be used to attach various materials to the shade.

Place mats made of vertical, thin bamboo strips can be wrapped around a drum or square shade and attached with tape to bind it to the top and bottom edges. Acrylic paint can also be used to paint or to stencil patterns onto a shade.

Decoupage

Paper cutouts can also be glued to a basic background (usually parchment or vellum) using a slightly water-diluted solution of white glue. After applying the glue to the shade and then pressing on the carefully cut design, a dampened sponge is used to smooth out air bubbles and adhere the edges. Applying a small amount of pressure with fingers can assure adhesion of stubborn edges. No air bubbles should be entrapped. They will blister later.

When the glue has dried, spray several coats of either matte or glossy Krylon® over the entire piece, allowing each layer to dry thoroughly before spraying the next coat.

A "matchstick" shade sporting a three-arm or spoke-washer fitting. Bamboo place mats can be used over a parchment or vellum shade. Edges are taped with a fibrous paper tape.

Fabric is laminated to a translucent plastic backing in this drum shade.

A decoupage-decorated shade with a bas-relief leopard by Virginia Merrill. The shade covering is lampshade vellum (Bienfang Co.). The three-dimensional effect is achieved by mixing French clay with Elmer's glue (tissues macerated with glue is an alternative), into a doughlike putty and shaping a paper cutout around the filler. Modeling tools are used to gently model the bas-relief form and press the stuffed paper leopard to the shade.

The Pierced and Cut Look

This effect can be achieved on a lampshade of parchment or vellum that has been readymade, but it is best to perform cutting and piercing operations while the paper lies flat. Plan a linear design, or one with an outline quality. Slice into the shade covering by partially cutting around contours. Another texture is achieved by piercing the shade with a needle to make fine holes.

Draw your outline design in pencil on the right and/or wrong side of the shade. With an X-acto knife, or another sharp, pointed blade, cut partially around the contours so that portions of the design, such as points of petals or leaves, can be pushed outward to project and curl in shallow relief, away from the shade.

A cut and pierced shade by Virginia Merrill. White edges actually stand away from the rest of the shade in shallow projections. Holes are pierced with a pin.
Courtesy: Virginia Merrill

Another version of a cut and pierced shade by Virginia Merrill. To enhance the shadow effect, watercolor is painted in gradations of gray on the back of the flowers and leaves. As light comes through the shade and its paper lining, the gray painting intensifies the shadow effects.
Courtesy: Virginia Merrill

Indicate details such as veins of leaves by piercing a series of holes at evenly spaced intervals. This should be done from front to back. When the lamp is turned on, the effect will be a drawing with light.

You may want to create an additional shadow effect by painting a gray value of shading on the underside of the shade, indicating curling and bending of forms such as petals and leaves, as well as depth and shadow. If you cut and pierce the design into your parchment that has been cut to accurate measure before the shade has been attached to top and bottom rings, you have an opportunity to line the shade on the underside (or back) with a thin "rice" paper, the kind used in block printing. This lining acts as a diffuser of light rays as they come from the cuts and tiny holes. Apply the lining paper with white glue along the top and bottom edges and along the seam edges. Make certain that the paper lies flat and that there are no wrinkles. When the glue has dried, the shade is ready to be attached to its rings. Binding will hide the narrow glue lines.

It is possible to line a ready-made shade that you cut and pierce, but the effect on the inside will not be as neat because the glue lines cannot be easily hidden.

Unusual Materials

LEATHER

A semirigid oak-tanned leather can be used to form an opaque shade. After the pattern is made, the leather is cut with either scissors or knife. Edges are sewn together with a leather stitching needle or by punching holes along edges of seams and using thonging or lacing to stitch seams together. (Various ways are shown in photos and diagram.)

After seams have been stitched, they should be pounded flat with a wooden mallet. This in effect acts as an iron would, pressing and flattening out seams.

Other edges can be glued with a rubber cement that has been applied to both faces and allowed to dry to a tacky state before edges are pressed together. Pounding with a mallet here, too, creases and flattens edges.

The top and bottom edges can be either glued or sewn around their respective rings.

MAKING A LEATHER SHADE

A paper pattern is temporarily attached to the leather with a temporary putty called Plasti-Tak. (When removed, the putty leaves no mark of any kind.)

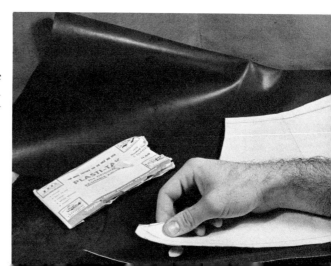

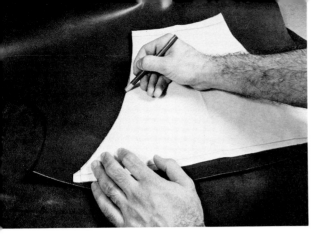

A grease (china-marking) pencil is used to outline the pattern on the leather.

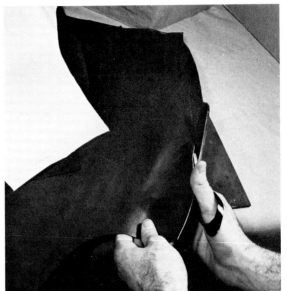

A pair of scissors is used to cut out the shade shapes. If the leather is too thick to be cut with scissors, then use a sharp knife.

Edges are skived on the back edges to thin out their thickness at the seams (where there is more than one layer of leather). A sharp knife is used to slice away a small amount of leather.

On the front of the leather, a marking or spacing wheel indicates where stitching is to go. A ruler and pin or awl can be used in the same manner; it just takes more time.

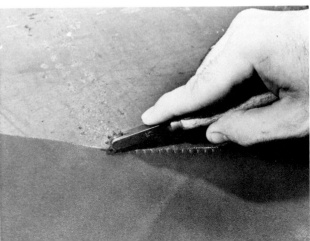

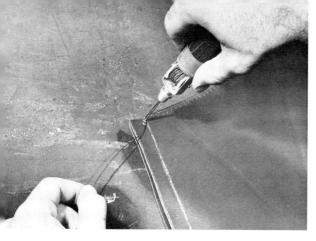

A lockstitch awl is used with matching waxed dark brown thread to sew the shade parts together. Here, the awl's needle, which contains thread from an enclosed bobbin, is pierced through the leather . . .

. . . and a free length of the same thread is threaded through the loop made by the needle. When the awl is withdrawn, the stitch is locked.

It is possible to create a similar effect using button thread that has been drawn through beeswax and sewing a combination stitch with a heavy-duty needle.

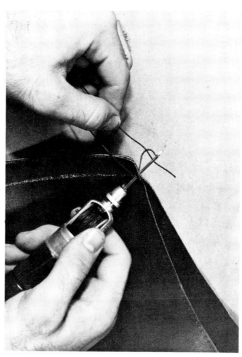

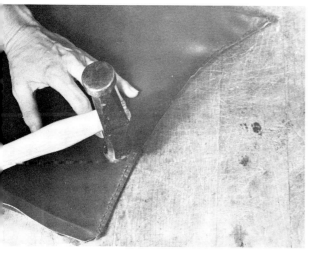

After the edges have been sewn, they are pounded with a hammer. In effect, this is like "ironing" the seam.

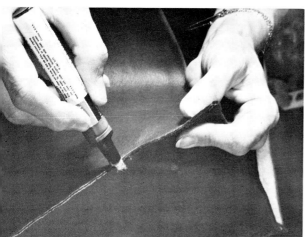

The raw edge is touched up with a dark brown marking pen.

The shade has now been sewn together and is ready for attachment to a ring, after the edges to be folded over are pounded with a hammer to form a crease. Rubber cement, coated on both edges to be glued, is brushed on.

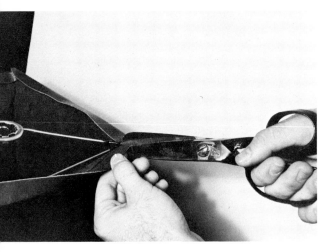

Before gluing down the top edges, a square is cut out of each corner to avoid bulky overlapping.

This photo shows both the corner cut out and a spider attachment that is connected at corners with tape to a wire hanger that had been bent to form the perimeter of the top "ring." This is a homemade solution that would have required an expensive custom order for a specially sized washer fitting.

The edges are sealed and the fixture is attached to the spider. The "homemade look" has been effectively covered with the leather. A bottom ring is optional. In this case it was sufficient to just fold under, glue, and pound the edges. There was sufficient thickness for the shade to lie flat and stiff.

The brown leather shade as a lamp in its environment, spotlighting a Balinese wooden sculpture.

COPPER

Copper foil, the kind used for making repoussé pictures, can also be formed into an attractive shade. Numerous effects can be achieved with copper because it is malleable and can be pounded with a mallet into the desired shape.

Parts are joined by overlapping and folding edges together just as one would make an apothecary fold when wrapping a package. Rivets can also be used, as can solder.

The surface can be left smooth, pounded into textures with a ball peen hammer, or scored into shallow repoussé patterns using a bone or metal tool that looks like a nutpick.

Pleated Shades

Paper sculpture forms are transferable to the lamp shade "department." Many kinds of folding and scoring techniques using heavy papers are applicable.

Pleated shades can also be made from some plastics or paper-clad fabrics. The only requirement in material selection is that the covering be semirigid, permit folding to a sharp crease, and respond to scoring with a knife or metal-edged tool.

Pleats are measured, folded in half, and then in half again.

A bone folder is used to define creases.

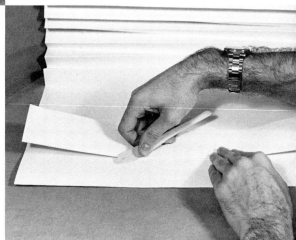

FORMING PLEATS

The accordion pleat is the simplest to fold evenly, though it requires precise scoring and folding. Depending on the shade proportions, pleat depth should range from 1/2 to 3/4 of an inch unless the pleating is to take on exaggerated proportions.

To achieve a sharply creased edge, it is necessary to score the crease line on the side opposite to the crease (behind it). Therefore, an accordion-pleated effect requires scoring alternately on the front surface and on the back. To score, use a bone folder (used in bookbinding) or a dull knife. Press it firmly alongside a straightedge (ruler) just hard enough to create an indented line. On the reverse side, fold the pleat. Then, run the bone folder blade along the edge to crease the pleat.

For the next operation, measure the 1/2- or 3/4-inch width on the front. Place your straightedge on the width indicator and score a second line. This time the fold will reverse itself. Continue measuring, scoring, folding, and creasing on alternate sides until the proper length has been reached.

ATTACHING OF RING TO CORD STRUNG THROUGH PLEATED SHADE COVER

After all pleats have been made, there are several alternatives to attaching them to the rings. One is shown here.

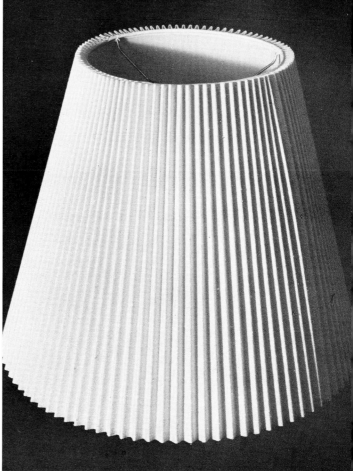

Another alternative is to tape and glue the pleats to the shade rings as shown here.

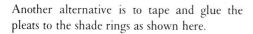

ATTACHING PLEATED SHADES

Pleated material can be attached and formed into a lampshade in three ways. One method is by gluing creased edges of the inside part onto a vellum shade. Use white glue and hold pleats in place by wrapping the piece with elastic—perhaps the elastic found on the inside of an old golf ball—around and around firmly but not so tightly as to indent the pleats.

A second method is to glue the top and bottom pleats to bound rings. Run a bead of white glue along the bottom ring. Tie an elastic cord around that area. Then repeat the operation along the top ring. Keep the pleats tied around or wrapped with elastic until the glue dries. Then adhere tape on the inside top and inside bottom edges, further bending the pleats to the ring.

A third method is to punch holes through the center of the pleats (toward the back edge) 3/4 of an inch from the upper edge and about 1 1/4 inches from the lower edge. Thread a cord or nylon filament through these holes and pull the string until both edges fit around their respective rings. Tie the cord securely. Then, thread more cord around the cord, running through each pleat and around the ring. The pleats will then be wrapped or tied to the rings. (See diagram.)

Pleating Variations

Accordion pleats are simple folds that result in a very attractive lampshade. Other striking results can be achieved by folding and refolding papers in several directions.

A heavy weight paper should be used and scored following patterns described in the diagrams and using accurate measurements. Score all lines shown in the diagram. Heavy lines indicate scoring on one side and broken lines denote scoring on the opposite side. Fold all vertical lines first, then all the diagonal lines—or one set of parallel lines first and then the next set of lines and so on. Crease the folds to achieve a sharp edge.

Proportions can be changed to suit your needs. All that is necessary is to follow the concept or design of folding.

Attachment of these forms to rings would proceed in the same manner as for the pleated shade described earlier.

◄ Paper pleats can be compounded into sculptural shade forms to be incorporated with a base or to exist as a shade-lamp. One alternative is shown here.

► The pattern for creating this form. Solid and dotted lines indicate that the paper should be scored and folded on opposite sides. After folding, the form is closed with paper tabs and white glue.

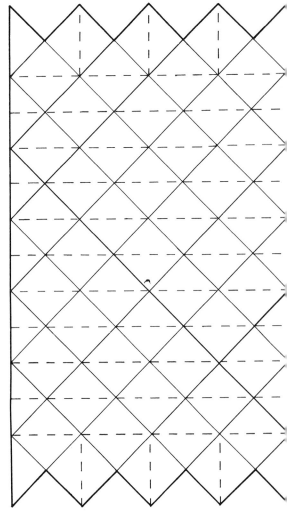

LAMPSHADE MAKING

This piece is composed of isosceles triangles . . .

. . . as shown in this folding diagram.

96 LAMPSHADE MAKING

The geometry of paper folding finds another application in this paper lampshade. Note that the two holes represent holes for the bulb, fixture, and light transmission.

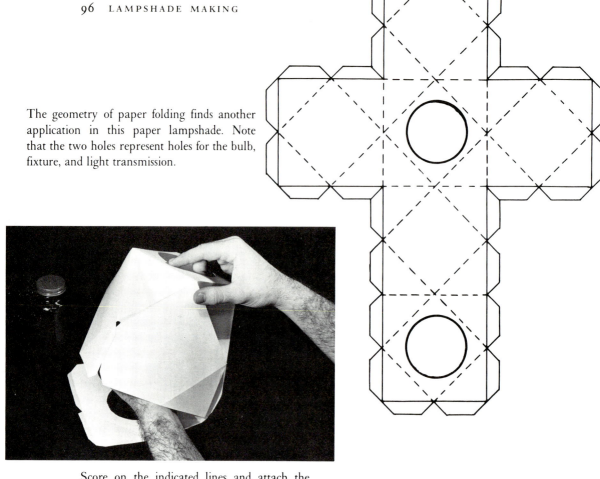

Score on the indicated lines and attach the tabs with white glue.

The completed lampshade can vary in size, depending upon how large or small the piece is scaled.

For another lamp as shade idea, start with an equilateral triangle. Score lines across each of the points, creating three more (really four if you count the center) equilateral triangles.

Fold the scored lines inward.

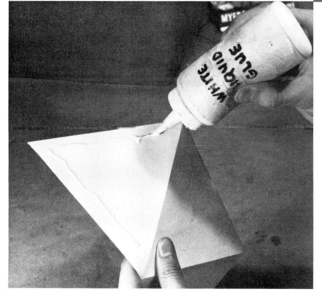

The three outside triangles are glued . . .

. . . and attached to other triangles with clothespins or large clips holding them in place . . .

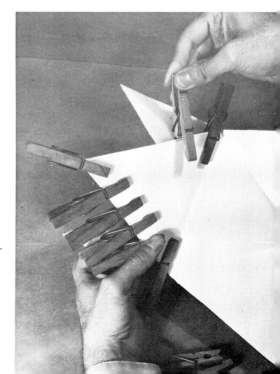

. . . until all fifteen triangles form this shade.

Another variation on paper folding is to pleat a sheet and then bend each pleat and make a sharp crease.

When the paper is opened, the creases will be visible in a zigzag across the entire width of the pleated paper. Follow the creases. The pleats on one side of this fold will have to be creased in the reverse of their original direction as you move along the paper. Gather in the paper as you fold.

Attach along the inside with a staple gun, matching both ends.

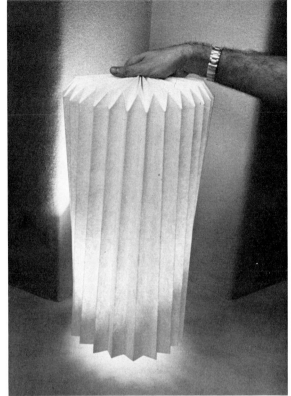

One possibility.

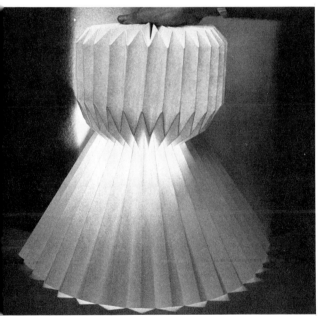

Other possibilities are achieved by making several folds. Angles can be changed to create different effects.

Here is another one.

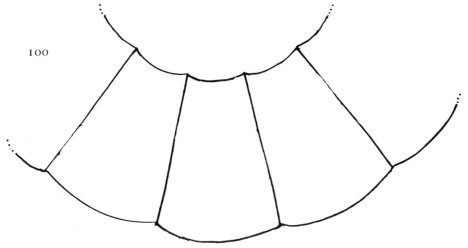

Partial Pattern Concept for Fluted Shade

Fluted Shades

Fluted shades are partially pleated with the outer section looping in an unbroken curve rather than being creased as in a pleated shade. The depth of fluting varies according to how extreme the curve of the arc is at top and bottom of each section. With a bit of experimentation, using scrap paper, initial patterns can be cut following the model shown here.

When the curve and depth of fluting have been achieved to your satisfaction, a heavy duty pattern should be cut. This will act as a template.

A fluted shade attached around another plain vellum shade.

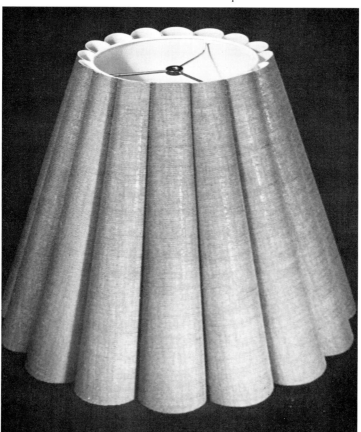

Trace around contours with edges juxtaposed until a full-sized pattern is rendered. A large sheet of paper is necessary because the length of paper template can be double or more compared with the length of a template of a typical conical shade. (See diagram.)

After the pattern has been cut using a straightedge and knife, score the pleated line (only on the outside of the shade cover), fold, and crease the pleat. Continue this operation all around the shade. Attach the shade to rings using one of the three methods described for attaching pleated shades. Glue ends with white glue or tape ends together with the same color paper tape.

Slipcover Shades

A slipcover shade works well over the ivy-training frame or hanging wire basket planter found in the gardening supply shop and used earlier to make a soft-covered shade. In this case, an opening must be cut in the top with a wire cutter to accommodate the light fixture.

The length of fabric to be used is the measurement around the circumference plus a seam allowance of one inch. The width is the distance from the top center (hole) to the bottom ring.

Seam the ends (width of fabric) and you will have a tube that should slide over the frame, snugly fitted at the bottom.

Stitch a 3-inch seam with a sewing machine or by hand at the bottom edge of the cylinder. Leave a small opening and thread elastic through the hole until it comes out from the other end (at the point where you began).

Fold under a 1-inch allowance at the top with heavy duty thread. Using a running stitch, gather the edge together until there is a 2- to 3-inch opening. Tie ends of the heavy duty thread tightly and tuck them into the seam. Now before tying the ends of the bottom elastic together, place the slipcover over the frame and pull the elastic so that the slipcover does not sag at the bottom. Tie the ends and tuck them away. There should be enough give to the elastic for you to be able to remove the slipcover. If seams were sewn, then the top gathers will have to be opened and released in order to remove the slipcover.

Wrapped Shades

Cord, raffia, wool yarn, ribbon, nylon monofilament, gift wrap cords, and so on can be wrapped around a shade frame that contains struts just by winding the material around and around until the entire frame has been filled in.

Shells or beads can be strung from the top ring to the bottom ring all around until the frame is covered. It is a heavy job—a weighty shade.

Capiz shells, or for that matter, boiled and shaped fish scales, can be sewn together in this manner . . .

. . . to create this lamp. From the Philippines.

A basket that has been lined with paper and electrified. From Bali, Indonesia.

Another basket, with Capiz shells hanging from the edge. From Lake Lanao, Mindanao, Philippines. Ways of converting baskets to shades can be seen in chapter 6.

Cut crystal beads are strung around a metal canopy shielding a 10-inch-diameter satin white glass globe. Beads and metal canopy are the same color. Lamp by Mobilite.

Courtesy: Mobilite

ABOUT HARPS

At the time when the shade frame is purchased or when the size of the shade is determined (if it is to be a table or floor lamp), a harp size and shape will have to be decided upon. As mentioned in chapter 2, two types are available—two-piece detachable and single-unit harps. But length and width vary to accommodate different heights of shades and sizes of bulbs. The harp fits below the socket and above the lamp base if it is the detachable type. The harp screws onto the socket housing directly if it is the single-unit type.

When fitting shade to lamp, the harp should never show below the bottom edge of the shade. If the shade is very long, then a recessed attachment at the top of the shade can meet the harp partway down.

If you are still not satisfied with the proportions, a shade riser 1 or 2 inches long can be used at the top of the harp. A finial is used here to attach the shade to the harp, with the finial screwing onto the male bolt at the top of the harp. It takes some experimenting and experience to figure out proportions. If you have the lamp base and shade frame on hand, somebody can hold the frame at the proper height in the air while you use a ruler to measure the harp size that you need.

4

The Portables: Table and Floor Lamps

Most often the lampmaker will have a specific location in mind for the lamp being designed. A particular place in a room and a definite purpose probably instigated the lamp building. But most lamps for home use are—and need to be—portable. Room layouts are changed; people move from one location to another. Many rooms do not have built-in connections in ceilings and walls, but only outlets for plug-in fixtures. Therefore, even though a lamp may be designed with a specific purpose in mind, it will likely be put to alternative uses.

The terms *portable* and *permanent*, therefore, have practical implications for lamp construction. The choice of materials, the location of switches and wiring, the durability of construction, are considerations influenced by whether a lamp is created to hang forever in one corner of a room or whether it will be shuffled periodically among different environments. There are, of course, ways of building in portability—even for ceiling and wall lamps. For example, with the use of one special canopy illustrated in chapter 2, a lamp can be unplugged from the ceiling as easily as one plugs and unplugs a table lamp from the wall outlet. For the most part, however, when we refer to a portable lamp, we usually think of a table, floor, or freestanding pole lamp.

Lamp parts and findings utilized in most table and floor lamps are discussed in chapter 2, as are various ways of wiring these fixtures. The essential construction techniques vary little from lamp to lamp. The application of the materials—wood, plastic, metal, stone, and glass—does vary significantly. Chapter 3 shows how to construct shades, which form a significant part of most table and floor lamp designs. And chapter 6 illustrates the use of a variety of found materials such as bottles, porcelain vases, and even sewer pipe in making portable lamps. This chapter mainly considers two of the most available and easily usable materials for the contemporary craftsman/lamp builder: wood and plastic. With a bit of practice in employing these staple materials, the lampmaker can execute a limitless number of designs.

WORKING WITH WOOD

Tools

Three levels of equipment are used in woodworking. *Basic hand tools*: screwdriver, hammer, knife, coping and crosscut saws, hand drill, plane, chisels, mallet, clamps and/or vise, awl, pliers, and ruler are on hand in many households. Few other tools are necessary. However, at the next level, *power tools* speed up operations considerably. An electric drill with a variety of attachments (router bits, sanding disks), a hand-held saber saw, an orbital sander, a circular saw, a miter box, are all excellent investments that take up little space. *Motor-driven equipment* such as planers, drill presses, routers, lathes, radial arm saws, band saws, disk sanders, and the like are wonderful to have, but clearly are *not* on the home lampmaker's scale. They can only be justified when a great deal of work is performed. Moreover, although it may take a little more time and effort and patience to use simple tools, the results are usually equivalent.

Basic Wood Processes

There are some fundamental ways of working wood. We can cut it, assemble it with glues (and less preferably with nails and screws), build up thickness of cut pieces, fasten two or more pieces together, apply thin layers (veneers) of one wood to another, and so on.

There are a few basic skills involved in performing each of these operations. One must be able to glue, apply veneer, sand, drill, plane, saw, carve, and bring the wood to a finished state. Mechanical operations will be illustrated as their applications arise throughout the chapter.

GLUING WOOD

The most attractive and enduring way of joining wood is to cut joints into the wood and join the parts with an adhesive. Among many glues there are two which work best: Weldwood Plastic Resin Adhesive® is a plastic that comes in powdered form and requires mixing with a small amount of water; Titebond® and the all-purpose white glues like Elmer's® and Sobo® (made of polyvinyl chloride and/or polyvinyl acetate) also work well with wood. Of these types, Weldwood adheres best. Two-part epoxies also serve well and are particularly advisable in some operations (as are some other plastic resin glues). But, generally, rely on Weldwood.

Since most lamps do not support great weight, simple butt and mitered joints are the most common—and practical—types of joints for lampmakers (especially when glue, the longest lasting means of attachment, is used). While fasteners like nails and screws work with some soft woods, they are the least satisfactory methods of joining wood, because they split hardwoods and may cause damage to any wood as the material ages.

VENEERING WOOD

Thin layers of wood—known as veneers—are available in rolled up strips from lumberyards and building supply stores. Most often used as shelf edging, veneer strips may be used to cover any surface. They are easily cut with a sharp knife and applied to lower grade woods (like plywood) with a white glue. They give the appearance of fine mahogany, oak, walnut, ebony, and so on. Veneers can be oiled, stained, and otherwise finished just like a solid block of wood. In fact, it is often difficult to distinguish a veneered piece from a solid block, if attention has been paid to grain direction during construction. Certainly, the final piece can be as attractive as solid wood.

TEXTURING WOOD

Texturing can be accomplished using *sandpaper*—by just sanding the surface until the desired roughness or smoothness is achieved; or one can use *Surform®* tools which contain tiny razor-sharp perforations that grate away small areas of wood. *Steel rasps* and *scrapers* can also be used to cut patterns into the wood surface, adding a decorative or distressed design to a lamp base. Emphasize textured areas by staining the wood.

Deeper textured patterns are achieved using *knives*, *chisels*, and *carving tools*. By cutting at an angle, light- and shadow-catching patterns can be achieved. Brace the wood before carving with C- or spring clamps. Set the chisel along the line marked for carving, starting with a cross-grain cut. Rap the head of the chisel with a wood mallet. This is the essential first step of chiseling. Next cut with the grain; you will see that, while it is easier to cut with the grain, if that first cross-grain cut had not been made then there would be little control over the area to be chiseled. Always chisel away from your body. Use larger chisels or carving tools to excise larger areas.

Special wood textures are achieved by *burning*. Heated brass or steel forms will scar wood in the old branding iron fashion to create patterns in the shape of the "iron." The surface of wood can be superficially burned using the flame of a propane torch, and charred wood may be scraped away with a wire brush, leaving areas of softer wood between the grain less charred. This creates a shallow relief that dramatizes the grain pattern.

FIGURINE TABLE LAMP

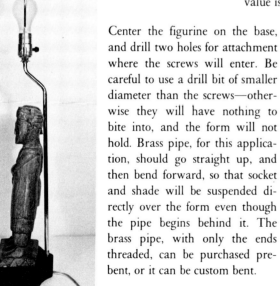

Wood is frequently used as the base of a lamp base—when objects are to be displayed such as converted vases and figurines. Cut a base from a block of wood and sand until smooth. Holes must then be drilled to accommodate threaded pipe, the lamp wire, and screws to hold the wood sculpture. Drill a 7/16-inch hole vertically through the base at its back edge to accommodate end-tapped brass pipe (1/8-IP). Drill another hole from the back of the base into the vertical pipe hole to thread in the wire. This hole should be made near to the bottom of the base so that the wire can escape cleanly. Near the center of the base, drill two more holes slightly smaller than the screws you will use to secure the sculpture. On the bottom side, use larger drill bits to countersink, or recess, the steel nut which will secure the threaded pipe and the screwheads. The base may be stained with a solution of black acrylic paint diluted with water. Brush the stain on, allow it to seep in for a few minutes, and wipe off the excess. Repeat this process several times, until the desired, muted value is realized.

Center the figurine on the base, and drill two holes for attachment where the screws will enter. Be careful to use a drill bit of smaller diameter than the screws—otherwise they will have nothing to bite into, and the form will not hold. Brass pipe, for this application, should go straight up, and then bend forward, so that socket and shade will be suspended directly over the form even though the pipe begins behind it. The brass pipe, with only the ends threaded, can be purchased pre-bent, or it can be custom bent.

The brass tube is attached to the base with a slip washer and hexagonal nut on the underside. Harp and socket attach to the threads at the top end. A coarse burlap shade (constructed in chapter 3) complements the primitive Taiwanese carving.

A CORK VENEER TABLE LAMP

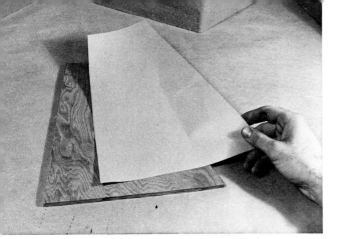

A shell of plywood, covered with dark cork and walnut veneer, produces a handsomely textured lamp. Cut four trapezoids of 1/4-inch plywood from the same pattern—to ensure that all sides will be identical.

Although it is best to use a glue to join fine wood parts, this pyramid-shaped base was attached with #18 wire brads (1/2 inch long). The brads offered a quick solution in a situation where strength and surface appearance (the wood will be covered) were not considerations.

The edges were butt joined and the nails driven through. A top, with a 7/16-inch hole in its center for the threaded pipe, must also be cut from 1/4-inch plywood. Set it aside for later attachment.

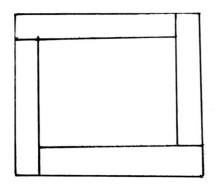
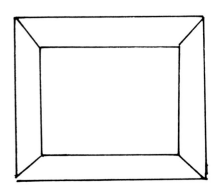

Two common types of joints in woodworking are the *butt joint* and the *miter joint*.

Using the same pattern, trace the shape of each trapezoidal side onto the back of decorative cork sheets. Add an extra margin equal to the thickness of the cork to one side. Cut the cork with a sharp knife, and use a mastic adhesive, like Ceramic Tile Adhesive® to adhere the cork to the plywood. Apply the adhesive evenly to the plywood using a tongue depressor or spatula. The cork does come with a sticky backing in many cases, but it is not strong enough in this application.

Press the cork to the adhesive-covered plywood. Apply pressure with your hands and with small weights. Allow the mastic to dry.

If the cork edges are a little uneven or stick out farther than the sides they butt against, use a sharp knife to trim away any excess. If the brittle cork has accidentally chipped away, apply a little mastic to the area, and press on a scrap of cork to cover. After the four sides of cork have been applied, nail the plywood top in place and cover it with a properly sized piece of cork. A hole must be drilled through the center of the cork as well. Trim the edges of the top piece of cork to the same angle as the cork sides.

A piece of badly marred pine may be used as a base for the pyramidal box—but it must somehow be disguised. Walnut veneer offers a good solution. Veneer comes as a roll of thin wood that can be cut with scissors or knife, bent to suit a shape, and glued in place. Before adding the veneer, however, all other operations on the pine base should be completed. A 7/16-inch hole must be drilled in the center of the wood, for the lamp pipe; a space must be drilled to countersink the washer and nut on the underside of the base; and a groove must be chiseled or routed from the center to an edge for the wire. To attach the veneer, cut four lengths, and trim the strips of wood at the corners using knife and ruler so that a 45° angle, like a mitered edge, is formed.

After cutting and folding the veneer, apply an even coat of white glue to the wood.

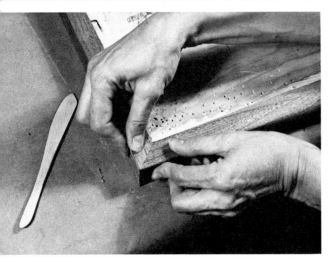

Allow the glue a minute to dry—so that it becomes sticky. Lay on the veneer, beginning at one corner. Press it down firmly, making certain that the "mitered" corners meet.

To ensure the veneer's bond to the base, use a burnishing tool, a smooth stick, or the back of a pair of scissors, to rub out any air pockets or unevenness in the veneer.

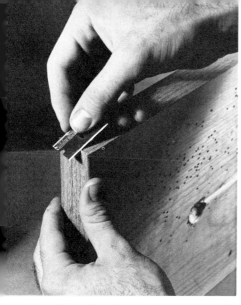

Trim the corners with a single-edged razor blade, making certain that there are no excess flaps of veneer. Also, cut away a notch of veneer where the chiseled channel for the wire meets the edge of the base.

Once the glue has dried completely, veneer may be finished like any other wood. Here, a walnut stain was applied with a brush. Excess is wiped off. To darken further, repeat the staining process several times. Note the recessed area in the center of the base to accommodate threaded pipe, washer and nut, and the chiseled channel which will carry the wiring from the center to the outer edge of the lamp base. Connect the cork-covered pyramid to the veneered pine base with the lamp backbone: threaded pipe (1/8-IP) and hexagonal nuts with slip washers. The threaded pipe should extend as high as you intend to place the socket, which will screw onto the pipe end. To mask the nut used at the top of the pyramid to hold it and the base tightly together, use a large check ring. Cover the threaded pipe, which extends from the top of the pyramid to the socket, with brass tubing. Attach harp, socket, and wire as discussed in chapter 2.

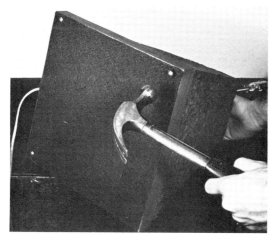

Cover the bottom of the lamp base with felt. The felt is adhered with Velverette® or white glue, and brass tacks are nailed into the corners of the base.

The lamp is topped with a matchstick shade in deep brown. One would never know that beneath that cork-veneer-brass exterior lies a soul of plywood-pine-and-steel-pipe.

Cork cubes are combined with glass globes of different sizes. A candelabra socket holds the light bulb, and the lamps have an on-off switch on the wire. *Courtesy: Mobilite*

A rectangular wooden base is covered with walnut and ebony veneers in a geometric design. The lamp is 26″ tall, with a 15-inch square white parchment shade. *Courtesy: George Kovacs Lighting, Inc.*

Mobilite combines 5-inch-diameter glass globes with hexagonal walnut bases of different heights. Hexagonal forms require miter angles of 60°. *Courtesy: Mobilite*

SHOJI STYLE LAMP

Die-cut Masonite panels, available in a variety of patterns, produce a handsome silhouette effect when covered with Japanese rice paper and lighted from behind. The Masonite, a dull brown in its unfinished state, may be spray-painted with any color. Here, glossy black was used. Lengths of molding, which will serve as stilts for the Masonite box, were cut and colored with a black alcohol-based stain.

Apply white liquid glue near the edges of the Masonite panel.

Starting at one corner, lay rice paper on the glued edges. Press the paper down carefully, avoiding wrinkles, and making certain that it is taut.

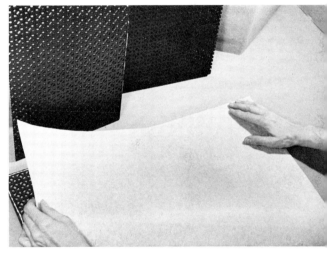

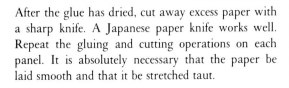

After the glue has dried, cut away excess paper with a sharp knife. A Japanese paper knife works well. Repeat the gluing and cutting operations on each panel. It is absolutely necessary that the paper be laid smooth and that it be stretched taut.

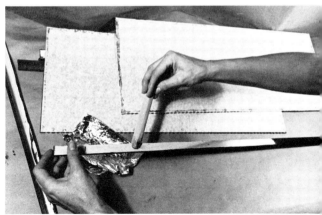

L-shaped molding strips—stained on the exposed portions—are glued to the panels with a two-part epoxy that hardens in five minutes. Do not apply too much of the adhesive, since the panels must be pressed tightly to the molding. If glue does squeeze out, wipe it away immediately.

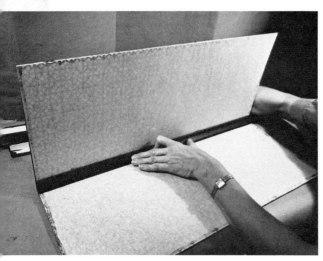

A molding strip with a right angle is pressed into the corner from the inside of the box. This strip is adhered with epoxy as well. Hold the sandwich —molding, panel, molding—tight until the epoxy hardens. Repeat the process in each corner.

The inside molding strips were cut so that they would stick out below the Masonite panels and act as legs, but the L-shaped molding used on the outside edges ends at the bottom of the panels. Stained pieces of 1-inch wood trim finish off the top and bottom edges. Spring clamps hold them in place while the epoxy hardens.

Another strip, of unpainted wood, is glued across the bottom of the Masonite box, to support the bulb socket. The wire is stapled along that strip and runs down one leg.

Because the lamp is tall, a tubular bulb, 12" long, is used to illuminate it. This type of bulb provides a very subtle light—especially when diffused by the panels and fine paper. The lamp may be built in larger sizes, too. This one, at 24", is table sized; larger forms might suit the floor, or be hung.

TURNED WOOD LAMPS

Turned wood forms, meant for use as table legs, are available at many lumber and home furnishings stores. Such turned forms are ideal for lamp bases—if they have holes drilled through the center to accommodate wiring. If they do not have holes, the forms may be drilled. One method of drilling is to cut the leg into shorter pieces—which may be drilled with conventional bits.

Each section is drilled, and the shorter units are glued together, creating a larger, drilled form. To determine what length each unit should be, double the length of your drill bit, since you will drill from each end of the piece. Often, due to the shape of the turned form, the slices will be invisible when the piece is reconstructed. Light sanding will also eradicate the line.

After drilling, glue the sections together with Weldwood®. Follow directions on the can for mixing.

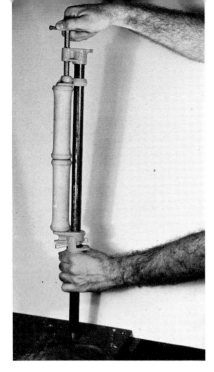

Apply some pressure, using weights or clamps. Allow the glue to harden for 24 hours. The turned piece—now with a hole through its center—may be wired, fastened to a base, and fitted with socket, harp, and shade.

Turned wood lamp on a dome-shaped metal base, with a round linen shade.

George Kovacs manufactures this 63-inch-tall floor lamp using lathe-turned western pine. It has a 21-inch-diameter pleated linen shade. *Courtesy: George Kovacs Lighting, Inc.*

A turned lamp base by Roy Child.

Akin to these turned forms is this bronze table lamp inspired by the traditional design of brass candlesticks. *Courtesy: Keystone Lamps*

This chunky lamp is both turned and carved out of solid western pine. It stands 18″ tall and has a 20-inch-diameter white parchment shade. Notice how the angle of the shade and base complement each other and how the simple shade suits the more complex base.

BENTWOOD LAMPS

Wood can be bent if it is steamed or boiled. When wood is both hot and wet, its plasticity increases. Boil the wood, allowing about one hour per inch of thickness. After softening, it should be bent as quickly as possible to minimize moisture and heat loss. Shape the wood over a firm surface such as a frame, mold, or jig. For extreme bends (as here) a tension strap will be necessary. For permanence, allow the wood to cool and dry while in its frame or mold—a process that may take as long as a week. The walnut laminate lamp shown here employs a 14-inch white Plexiglas shade. The lamp is mounted on a wooden base using a dowel joint. The shade adjusts from 46″ to 54″. *Courtesy: Koch & Lowy, Inc.*

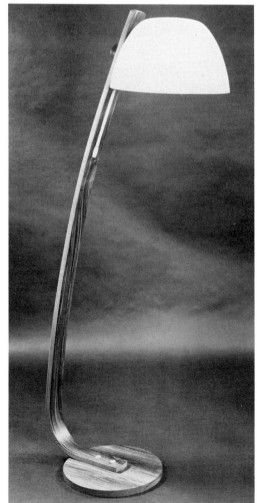

Above, left:
A chrome shower curtain rod was cut into random lengths, fitted with tiny sockets, and topped with gray ball-shaped decorator bulbs. The whole is mounted on a walnut base. By Lee S. Newman.

Above, right:
Stained glass "Tiffany"-style ceiling lamp by Jack Cushen.

A cork-clad plywood pyramid-shaped base mounted on pine that has been trimmed with walnut veneer. The hard shade consists of thin strips of wood (that look like applicator sticks) over a white plastic.

Tubular knitted acrylic is stretched at intervals around shade rings of different diameters. Lighting consists of a string of decorator bulbs.

A nest of acrylic rods shades three ball-shaped bulbs that hang from a ceiling fixture. By Thelma R. Newman.

A potichomania base in a seashell theme by John Campbell and Lewis Morrow.

Transparent Poly-Mosaics are adhered to a clear acrylic tube. It sits on an X-shaped walnut base.

Scraps of walnut form a box around a black acrylic liner that can also function as a lamp base. The hard shade is made of natural linen laminated to plastic and trimmed with grosgrain ribbon.

A leather shade in a "pagoda" shape accents an antique Balinese sculpture.

A basket as shade, lined with paper. From Bali, Indonesia.

Twine, saturated with glue, wrapped around a balloon, formed this sphere-shaped shell. A rattan chain suspends the lamp and its cord from a cantilevered wooden wall bracket.

A Taiwanese primitive sculpture is mounted on a wooden base. The self-trim burlap oval shade and lamp housing are attached via a brass tube that connects to the base.

An oriental paper lantern sits on a base made up of acrylic rods that are glued to an acrylic disk, edged with an acrylic ring.

A square tube of milky acrylic is fitted near the base with a square that will house the socket. A strip of metalized Mylar defines the division between the lighted and unlighted areas.

Acetate attached to rings by means of black plastic lacing sewn in a whipstitch. The lamp is suspended over a simple wooden base.

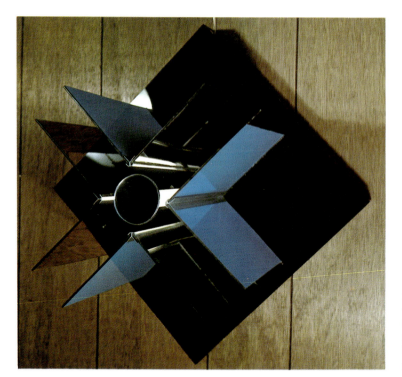

A wall sculpture-lamp by Lee S. Newman consisting of four right-angle shapes of acrylic mirror, backed with a blue finish. They are attached to a black acrylic box. An acrylic mirror disk is clamped to the showcase bulb.

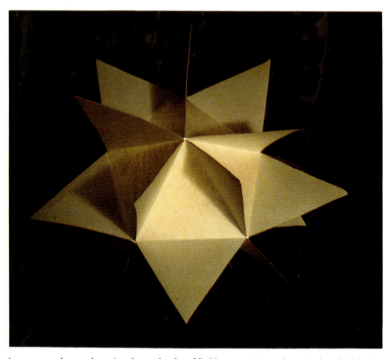

A paper sculpture hanging lamp by Jay H. Newman is made entirely of folded triangles.

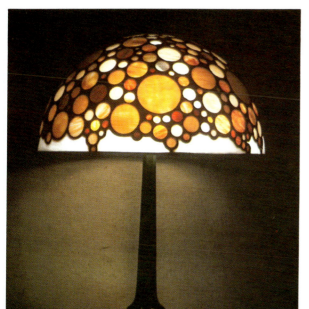

A table lamp with stained glass shade by Jack Cushen.

A decoupage lamp and shade by the late Lydia Irwin.

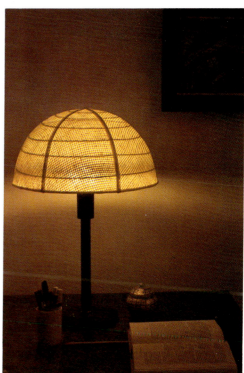

An ivy training frame, purchased from a garden supply center, formed the base for this soft shade made of a mesh weave natural linen. The shade rests on a reflector container that houses a bulb.

A "crystal" chandelier consisting of clear styrene disposable drinking cups. A large round bulb is enclosed.

Faceted acrylic domes attach with clips to form a sphere. Light refracts in crystallike images.

Plaster of Paris hands hold the acrylic mirror tube that contains a fluorescent lamp. Lamp by Lee S. Newman.

A POLE LAMP

Philippine mahogany was chosen for this pole lamp because it takes on a warm, rich color when oiled. Although it is a hardwood, it is not as dense as some others, and, consequently, fabrication is somewhat easier. The pole lamp was designed as an elongated box to house wiring and fixture connections. To construct the pole, first cut four box sides roughly to size. The width should be exact, but the length will be trimmed later. Do not worry about the exterior surfaces—except to check for very rough areas which will need to be sanded smooth later. The wood is joined by gluing edges as they butt together with Weldwood Plastic Resin Glue®. This adhesive must be carefully mixed according to the instructions on the can. A thin bead of glue is applied along the surface to be joined, and the pieces are clamped together firmly using wood clamps (as illustrated) or using C-clamps (with cloth beneath the metal to protect the wood). Do not worry if the glue squeezes out of the joint; this is normal. It is called "ooze out" and will be sanded away later. Weldwood should be allowed to dry overnight, or longer if humidity or cold air interferes with proper curing.

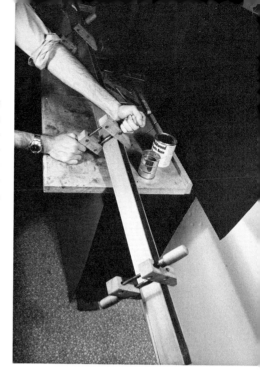

After the glue has dried, the exterior must be sanded to a smooth finish. This may be done by hand, or with a belt or disc sander. Spacers are placed inside to keep the box solidly braced during sanding. One side will be screwed on later. This allows for subsequent removal and rewiring. That side is sanded separately.

After sanding, and while the spacers remain inside the box to maintain proper positioning, the top side is laid on and holes are drilled for screws. Do not attempt to insert screws without first drilling holes a size smaller than the screw. The pressure of too many screws at the edge of the wood could easily cause it to crack. Screws are countersunk—by drilling the top of the hole with a bit the diameter of the screwhead, so that the screws will lie flush when finally inserted.

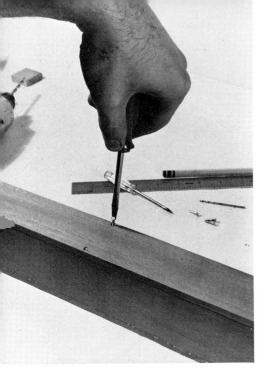

As holes are drilled, screws (drawn over wax or soap to ease entry) are driven in every six inches to keep the top aligned.

Carpet-covered tacks, available in housewares departments to protect carpets and furniture, are driven into each end to protect floor and ceiling. Notice that the top and bottom pieces are screwed in, to allow for removal if necessary.

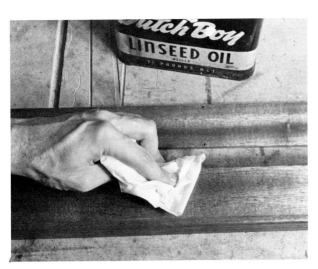

After all mechanical operations have been completed, the wood is finished by applying linseed or teak oil to the surface. Oils not only bring out the natural color and the grain, but preserve the wood as well. Give the wood several coats on the first day, periodically, as the wood absorbs oil. Then, after the lamp has been completed, oil it once a week for a month, and once a month after that. Before constructing external parts, wipe away excess oil.

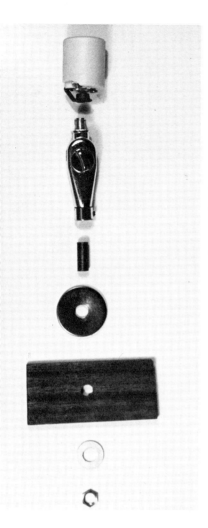

This shows the exploded socket/swivel findings. The swivel is probably the most important part of this unit, since its quality will determine how well the individual light will function and focus. While swivels tend to be somewhat more expensive than other lamp parts, do not skimp on them. At the bottom of this photograph, below the wood, lie one *slip washer* and one *hexagonal steel nut*. On the outside, in ascending order, are: a decorative *check ring*, which will cover the hole, protect the wood, and add a brass highlight to the lamp; the *threaded steel pipe* which will become a hidden connector for the swivel through the wood to the nut at the back; the *swivel*, one of many designs which will allow the sockets and lampshades to twist and turn; the porcelain *lamp socket*.

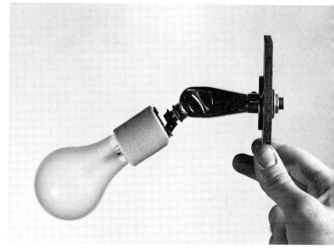

This is the entire assembly without the lampshade. To attach the unit to the pole box, drill a 7/16-inch hole in the wood to accommodate the threaded steel pipe.

In this case, lampshades were constructed from mirror-coated acrylic. Here, the tube of plastic is being cut by rotating it under the stationary blade of a radial arm saw. Other materials may be used to create shades for a pole lamp as well. In addition to traditional hard-covered shades in a cylindrical shade, consider the possibility of commercial metal shades, tin cans, glass parts, baskets.

The mirrored cylinders are given perforated tops of black acrylic (for ventilation), which are attached with methylene chloride. Holes were drilled in the disks in concentric circles using a drill bit meant for plastics.

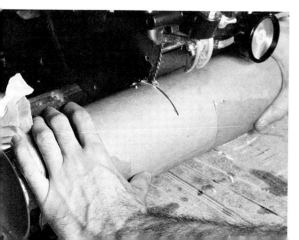

Three such lamp assemblies were connected to the pole to complete a lamp. A rotary switch was installed in the pole at thigh level for easy access, whether standing or sitting. The lamp itself is held in place through the tension created by another smaller pole that slides within the large unit. Brass rods are passed through the smaller pole just below the end of the main unit, holding the larger pole in place while tension on the brass rod maintains the pressure needed against the ceiling and floor, respectively, to keep the lamp vertical. The "rod system" solves the problem admirably and eliminates the need for springs or other mechanical devices. Because acrylic mirror distorts when exposed to heat, low wattage blubs must be used. If more light is required, use glass shades instead.

FINISHING WOOD

Sanding

Before applying any special finishes to wood, the surface should be sanded well (unless a rough texture is desired). Start with a rough grit, open coat garnet paper, like #80, and work your way to #220 or #400. Finer surface finishing can be achieved with #0000 steel wool. After sanding, wipe away dust with tack cloth (a waxy cloth to which sawdust sticks) before applying the finish.

Staining

Several kinds of transparent stains are available. There are *water stains* that are water soluble. These should not be used on tool-marked surfaces or on pieces where end grain is predominant. They raise the grain, creating a spotty effect.

Alcohol stains, such as the NGR (nongrain-raising) types, are easy to apply and are fast drying, but can leave a blotchy look if there are soft spots in the wood.

Oil stains provide an even coating. They are dyes which are dissolved in oil. These tend to be less permanent than the alcohol stains.

Pigmented wiping stains are diluted paint products which are applied by brushing or wiping with a soft cloth.

To avoid excessive stain penetration of any of these stains (particularly with end-grain woods) apply a dilute solution of shellac (diluted with alcohol) before staining.

Clear Finishes

Clear finishes simultaneously preserve the wood and reveal the natural beauty of the surface. Of the clear finishes, linseed oil or teak oil are best. Oils can be wiped on as they come from the can, or heated to about 80° F and applied. After removing all dust with a tack cloth, apply the oil with a brush or rag and allow it to soak in. For best results, apply several coats during the first day, and then one a week for a month. Each time, dry and rub the surface to spread and remove excess oil.

Petroleum jelly imparts a silky appearance to wood and does not produce the greasy surface one might anticipate.

Hard carnuba wax (used for floors) can be heated to its melting point and applied in a warm room. Allow the wax to penetrate. Wait an hour and then buff the wood with a clean shoeshine brush.

Shellac is one of the oldest and best clear finishes for wood. Use it when fresh and apply it on a dry day. (Stale shellac stays tacky forever.) Cut the shellac by mixing about four parts alcohol to three parts of three-pound cut shellac. The mixture is usually brushed on and allowed to dry thoroughly before additional coats are applied. Sanding between coats improves the final finish.

Many good varnishes which provide excellent finishes are on the market. They are applied by brush the same way as shellac. Follow directions on the can.

PLASTICS TECHNIQUES

Acrylic has many applications in lampmaking. Recognized most often by the trade names of Lucite® (Du Pont), Plexiglas® (Rohm and Haas), Acrylite® (American Cyanamid), and Perspex® (Imperial Chemical Industries, Ltd.), this plastic is outstanding for crafts use because of its versatility and the ease with which it can be formed and machined.

Acrylic weighs approximately half as much as glass of comparable size and thickness and has optical clarity (about 92 percent light transmission) that rivals that of less flexible, more fragile glass. Shrinkage and decomposition are minimal.

Acrylic can be sawed, drilled, tapped, sanded, polished, and otherwise machined, cemented, etched, or carved. Standard metalshop and woodworking tools are sufficient to execute most procedures.

Acrylic also can be heated and shaped while it is hot (softening between 240° and 340° F). When it cools, the new shape will be retained. If reheated, though, this assumed shape will be lost. This is characteristic of all "thermoplastics." Thermoplastics have a "memory" which, under heat, causes them to return to their original flat (or other manufactured) form. Heating acrylic (for bending or shaping) should be done carefully and evenly because, if overheated, shrinkage, scorching, and bubbling may result.

Acrylic resists corrosion by most household chemicals, but turpentine, benzene, lacquer thinner, acetone, ketones, along with some solvents, will attack acrylic, causing it to craze or cloud.

Acrylic is best known in sheet form which usually is four feet by eight feet. In sheet form it comes in thicknesses from 1/16 of an inch up to several inches. Patterned, textured, even mirrored and richly colored acrylic sheets are manufactured, and more varieties are continually being offered.

Blocks of acrylic can be obtained in chunks up to 1 x 2 x 3 feet. Sheets can be laminated into blocks as well. Extruded and cast rods and tubes (round, square, spiral) can be found in diameters ranging from 1/16 of an inch to 18 inches.

Marking and Sawing Acrylic

To mark the plastic before sawing, acrylic sheets and blocks are sold with a protective paper or plastic covering which can be peeled off. Marks can be made in pencil or pen on this temporary surface. If it is necessary to make lines directly on the acrylic, use a grease pencil, since it can be wiped away with a soft cloth afterward.

Acrylic sheets, rods, and blocks are easily cut using any saw made for wood- or metalworking. When using power equipment in cutting acrylic sheet, the plastic should be fed through the saw slowly to cut down on excessive friction and heat. To achieve a straight cut, a circular saw is best. Jig or saber types are most effective when cutting small curves or intricate patterns. Band saws are good for cutting larger curves.

Metal-cutting blades are best for acrylic since they deliver the cleanest, sharpest results. If using a circular saw, the blade should preferably be carbide-tipped, with teeth of uniform length. As with any power tool, the proper shop precautions should always be taken, including the use of protective glasses. If you do not have the tools for cutting sheet acrylic, many plastics suppliers can cut the acrylic to order, or sell it precut in manageable sizes.

Drilling and Tapping

When drilling into acrylic, use metalwork bits or the drill bits now available designed specifically for plastics (having a zero rake angle that prevents the bit from catching and fracturing the plastic). Tap acrylic just as you would tap metal. When drilling or tapping, apply mild soapy water to help reduce friction and lubricate.

Finishing an Acrylic Edge

SCRAPING AND SANDING

With hand tools one can produce a professional-looking "dressed acrylic edge." The result is an edge that will not have a high gloss, but will be smooth, translucent, and matte. Use a sharp woodworking scraper and wet-or-dry sandpaper. The scraper acts much as a plane does on wood, smoothing a normally coarse machined edge and removing any deep scratches from the acrylic. After scraping, wet sanding from coarse to fine (a recommended progression would include grits 150, 220, and 400) completes the dressed edge. Because it is not a glossy, clear edge, the dressed edge will "pipe" light. One of the properties of acrylics is that when they are scratched or sanded, light will shine through the marks. If polished to clarity, there will be no such "piping" of light. This is a useful property, and is discussed in relation to sculptural lamp designs in chapter 8.

TABLE LAMP OF ACRYLIC AND WOOD SCRAPS

A black acrylic box forms the base for a wood scrap lamp. Acrylic is readily cut with a radial arm or circular saw. Here, sheet is being ripped into the straight-edged pieces which will form the black acrylic box. Always leave the protective paper on the acrylic during machining operations.

To create a smooth, matte finish on the edge of acrylic, put the plastic in a vise and scrape off any rough spots and deep scratches using a wood scraper.

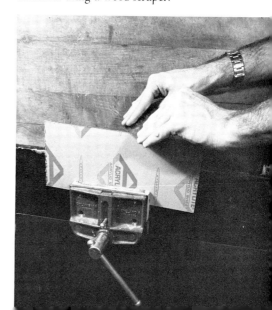

Sand the scraped edge, beginning with #220 wet-or-dry sandpaper. Wrap the sandpaper around a block of wood to guarantee flat sanding. Wet the sandpaper slightly, and progress to finer and finer grits until #400 sandpaper is reached. This produces a dressed edge.

For a glossy edge, a buffing wheel must be used. After scraping and rough sanding, apply stick wax and compound to the buff.

Polish the edge of the plastic, moving it back and forth across the buffing wheel. Do not hold the plastic in any one spot too long, or else the wheel will "dent" the edge, making it uneven. Polish only those edges which will not be glued. The dressed edge provides a better gluing surface.

POLISHING AND BUFFING

If you have access to a two-wheel electric buffer, a soft, fairly loose buff 10" in diameter at 2,000 surface feet per minute will give the best results with acrylic. One buff should be used with white tripoli compound to remove any scratches. Only light pressure is required to clear the acrylic surface of shallow marks or to polish a dressed edge to clarity. Very deep scratches should be scraped and sanded before polishing. Leave the second buffing wheel clean, to buff excess compound from the acrylic and deliver an even glossier sheen.

If using a faster, harder buff, apply less pressure to avoid cutting into, or denting the surface of the plastic from the heat generated.

Cementing Acrylic

A strong, almost invisible bond can be made by cementing acrylic to other acrylic, depending on the type of cement used. Some of the best results can be obtained by using a solvent cement like ethylene dichloride or methylene chloride. To bond an edge of acrylic to another piece (as in making an acrylic box), butt the ends, and apply masking tape to hold the pieces tightly against each other. With a fine brush, or syringe, touch the solvent cement to the butted joint. Cement placed at the edge of a joint will spread along the entire butted surface under light pressure, due to capillary action. The solvent will dry in three to ten minutes, be safe to machine in four hours, and will result in a powerful, cohesive bond.

An even stronger bond can be achieved by letting the solvent soak into the acrylic before applying pressure.

The best bonding cement for acrylic, however, is PS-30®. This exceptionally strong acrylic adhesive may be used to create watertight joints. It comes in two parts: an acrylic monomer (a resin called polymethyl methacrylate) and a catalyst (a hardener for the resin). The proportion of catalyst (compound B) to resin (compound A) is 5 parts to 95 parts, by weight. That is, if mixing 95 grams of resin, you should add 5 grams of catalyst. Stir them well without entrapping air. Let the mixture stand for five minutes so that any air bubbles can dissipate. Then apply the PS-30 mixture with a tongue depressor to the acrylic surface to be bonded. It sets in an hour and is hard in twenty-four hours. Resin and catalyst should be stored in a refrigerator. The resin has a very long shelf-life, but the catalyst, if kept too long, turns yellow. PS-30 may be used to bond butted edges of acrylic and to laminate sheet upon sheet of the plastic, as well.

To cement acrylic to other materials, such as metal, requires some experimentation. For most unusual bonding jobs, products such as two-part epoxy and Crazy Glue® could be tried, often with success.

The acrylic is stripped of its paper protection and cleaned with a soft rag and alcohol. To butt-join the acrylic, use methylene chloride. Butt the acrylic together, apply a little pressure, and draw the solvent cement along the joint. Use a brush or a plastic syringe to apply this cement. The liquid enters the joint by capillary action. Methylene chloride hardens in a minute to create a firm bond, but it does not completely harden for several hours. Repeat this process for the other edges, to complete the box. A top to the box should be cut and dressed and polished and cemented on as well. Before attaching it, however, drill a 7/16-inch hole in its center to accommodate the threaded steel pipe.

A walnut base was cut and sanded smooth. Working on the bottom, diagonal lines were drawn to determine the center, and a 7/16-inch hole was drilled to accommodate the lamp pipe.

Another hole was drilled near the edge for a single pole rotary switch.

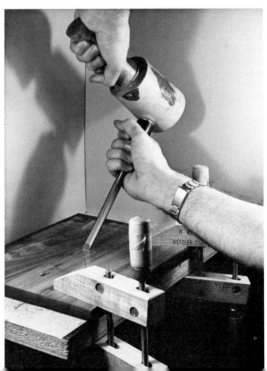

Using a chisel and wooden mallet, a recessed area was carved around the center hole to contain the washer and nut which will hold the threaded pipe. Parallel clamps secure the wood during chiseling. Begin chiseling by making "stop cuts" straight down and across the grain of the wood, then chisel with the grain. Without stop cuts the wood splits. Always chisel away from your body, and keep fingers and appendages outside of striking range.

A channel must be chiseled from the center hole to one edge to accept the lamp wire. Manual chisels may be used, or a router attachment which fits into most hand drills may be used (as here). A space was also chiseled out to accept the switch.

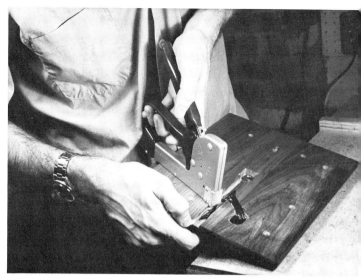

The lamp may then be wired. After splicing, soldering, and taping the wires, extra large staples were shot into the wood to hold the wires in place. Be careful not to pierce the insulation of the wires.

Threaded pipe cut to the proper length is attached to the base with slip washers and hexagonal nuts on each side. The acrylic box slips over the pipe, which fits through the hole in the box top's center. The process of attaching the pipe, masking the exposed portion with brass tubing, and wiring the rest of the lamp are shown in chapter 2.

The lamp may be considered complete as it is: black acrylic on a walnut base with a textured cylindrical shade. But you don't have to stop here!

Scraps of walnut may be combined into panels, glued together, and mounted around the acrylic form to create a very appealing mixed-media lamp. The scraps are first carefully sanded to smooth all surfaces. Arrange scraps of wood so that they form four panels similar in size to the sides of the acrylic box. Aluminum trays were used here, because you can glue directly on them and the panels will still separate.

Weldwood Plastic Resin Glue® is mixed in a tin can according to instructions. The powder is first spooned into the can; water is added gradually . . .

. . . and the mixture is stirred until the glue is smooth and properly syrupy.

Brush the glue onto the surfaces to be adhered.

You will need to employ a variety of clamps to hold the panel pieces together, because of the many different angles involved. Wipe away any "ooze out" of glue with a clean cloth. The pieces were presanded because it would be much too difficult to sand the panels afterward, and all excess glue was removed to avoid the need for further sanding afterward.

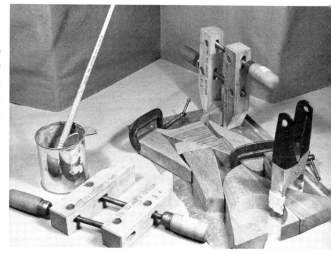

Weldwood should be allowed to cure overnight. Then sand away any roughness remaining (due to missed areas or ooze out), and apply linseed oil to bring out the deep natural color of the walnut. Brush it on, allow it to soak in, and wipe the excess away. Repeat this process several times during the first day, and once or twice during the next few weeks.

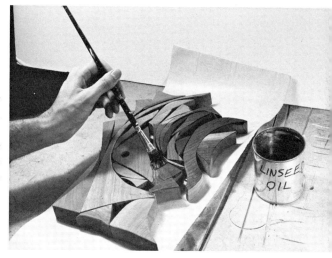

Holes must be drilled through the wooden base to accommodate screws that will hold the panels upright. Drill into the panels a little bit as well, and be certain to countersink the screwheads.

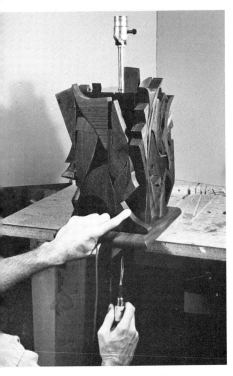

The acrylic box sets off the walnut. Screw the panels on tightly, but be gentle with them, because the pressure of the screws can crack the wood too.

Spread Velverette® or another all-purpose white glue evenly over the base to attach felt.

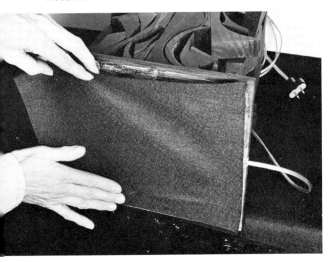

Gently apply the felt beginning in one corner and working back. Keep the surface taut and free of air pockets as you smooth on the fabric.

Trim away excess felt with a sharp knife.

A simple, rectangular linen shade trimmed in grosgrain ribbon complements the walnut and acrylic lamp base. The shade construction is illustrated in chapter 3.

Hurricane lamps are really sleeves of glass or plastic which originally fit around a flame to keep it from blowing out in the wind and rain. The form has been adapted to decorative purposes, however, as this candlelit smoked acrylic hexagon demonstrates. Smoked acrylic acts as a diffuser, but allows the forms within to shine through as well as the light. Regular geometric shapes with more than four sides must be mitered at different angles, depending upon how many sides there are. This unit consists of six sides, each 1/8 x 3 3/4 x 12". Each edge was mitered at a 30° angle, polished, and cemented.

ACRYLIC HURRICANE LAMP

Before gluing, the protective paper is peeled back from the edges. If solvent cement seeps between the plastic and the covering, the paper becomes very difficult to remove, and the acrylic becomes scarred.

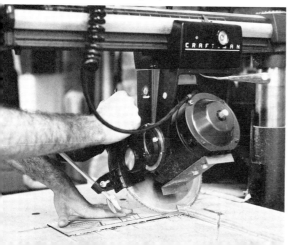

Angle clamps, set at 30°, hold the sections firmly in place for gluing.

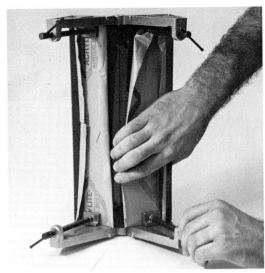

Attach a second clamp at the other end so that the entire end butts precisely. Tape may also be used to brace the mitered edges, but this method is far inferior to the use of angle clamps.

Apply solvent cement along the joint, and allow the pieces to set for several minutes. Follow the same procedure at each joint. If desired, you may first cement three pairs of sides together, and proceed from there.

Clear acrylic, 3/8 of an inch thick, will form the base. Trace around the assembled sleeve to determine the precise shape. Cut the piece, polish the edges, and drill a 7/16-inch hole in its center to accommodate wiring.

A groove is routed around the outside edge of the base to provide support for the hexagonal sleeve. To avoid some routing, you may cement a hexagon of acrylic a bit smaller than the sleeve's inside perimeter to the base. This will also securely hold the lampshade.

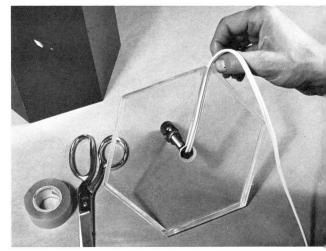

Some routing will be necessary, however, since a channel must be made in the bottom side of the base to carry wiring from the center to the edge. The bottom nut is set in a recessed hole, and the cord is taped into place with clear Mylar tape.

A 3-inch steel nipple supports the candelabrum socket. The cardboard sheath which comes with the socket is an important insulator, and it should be slipped over the socket.

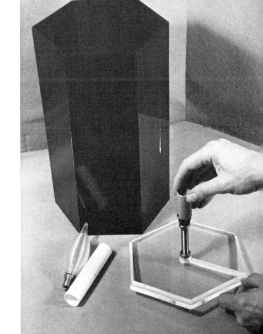

A white plastic cylinder—made for just this purpose—fits over the socket, insulation, and nipple. It imitates a wax taper.

Install a candle-shaped bulb, and lower the hexagonal form (after cleaning it with alcohol) onto the base.

This lamp glows from within.

This elegant table lamp consists of a clear acrylic box, pierced by a chrome pipe which suspends a 22-inch white parchment shade. *Courtesy: George Kovacs Lighting, Inc.*

These lighting fixtures employ smoked and opaque white acrylic. Note the use of acrylic's diffusing *and* reflecting properties in this series of vanity lights. *Courtesy: Richard Morgenthau Co., Inc.*

Cut, glued, and polished acrylic suspends metal orb lights in this ingenious design by Neal Small. The 14-inch desk lamp holds a reflector bulb. *Courtesy: Neal Small Designs, Inc.*

ACRYLIC DIFFUSER OF EXTRUDED TUBE

Acrylic is available in a range of extruded shapes such as round and square tubes. These materials alone allow many simple, effective lamp designs to be executed. Four-inch-square tube saves the trouble of cutting and gluing individual pieces of acrylic: all that needs to be done is to cut the tube, or to ask the supplier to cut it to size. A base for the socket will be suspended inside. To determine the size, trace around the inside of the tube.

Cut out the pattern and use it as a template. In this case, black acrylic was chosen as the socket base, in order to cut off all light in the lower portion of the lamp.

Drill the base so that it will permit the insertion of two bolts which will secure the sockets. Two more holes must be drilled to allow passage of the wires.

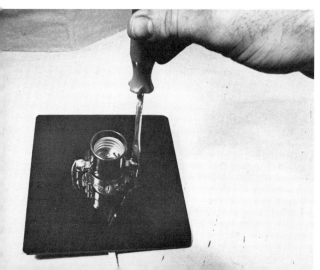

After bolting the socket down, pass the wires up from below, strip them of insulation, and attach them to the screw terminals as explained in chapter 2.

Strips of acrylic will support the platform within the tube. The first step is to measure down an equal distance on all four sides and mark the point. Then place the strips of acrylic along the marking and secure them temporarily with masking tape.

Draw a brush laden with solvent cement, methylene di-chloride, along the juncture of acrylic strip and tube. When the joint is firm, remove the tape.

A notch was routed out of the bottom edge to allow the wire to exit. But a hole would have served the same function.

Light does not penetrate the opaque black acrylic base. The point of transition is accentuated with silver metallized Mylar tape. An in-line switch completes this simple, functional lamp.

JAPANESE SHADE WITH LIGHT PIPING ACRYLIC RODS

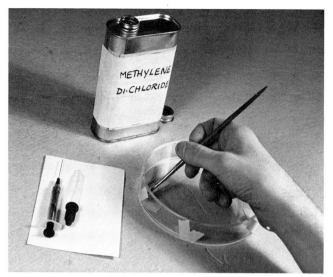

This mood lamp capitalizes upon the light piping qualities of acrylic rods. A ring of acrylic, cut from acrylic tube, is glued to a circle of 1/8-inch sheet. To facilitate gluing, tape ring and circle together, and allow solvent cement to flow into the joint. Drill a 7/16-inch hole in the center of the circle to accommodate a steel nipple which will support a socket.

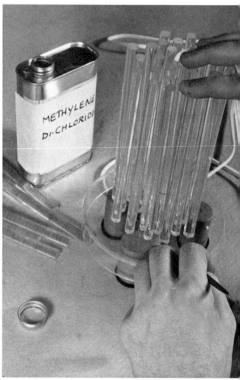

Extruded acrylic, available in diameters from 1/16 inch to several inches, may be cut, scraped, sanded. and polished just like any other acrylic. In this case, 1/2-inch rods with dressed ends are glued to the bottom of the disk.

A socket, supported by a steel nipple, extends through the center of the form, and a paper lantern set into the plastic dish serves as the shade.

Heating and Forming Acrylic

As mentioned earlier, acrylic sheets become soft and flexible when heated at a temperature between 240° and 340° F. After a few minutes of heating they can be shaped. Once cool, the shape created while the acrylic was hot will be retained, unless the plastic is reheated.

A regular kitchen oven, for heating large areas of acrylic, or a "strip heater," for bending acrylic along straight lines, is most often used by craftsmen to heat-form acrylics. Using asbestos or other protective gloves, hot pliable acrylic sheets can be handled and held in three-dimensional shapes until they cool.

When heating the acrylic in the oven (preferably on a piece of soft flannel cloth), be careful not to exceed the 240°-340° F temperature range. If left in the oven or over the strip heater for more than a few minutes, the acrylic will begin to scorch and bubble, leaving a textured surface which, depending upon your intentions, may be desirable or undesirable. The heating acrylic should never be left unattended. (There is little possibility of fire, however, since most ovens only heat to 450° F, and acrylic ignites at 700° F.)

Hot acrylic can be made to hold any shape by using spring or C-clamps, by (gloved) hand holding it, or by sagging it over or into other objects until cool. Cooling can be speeded up by putting the form in cool water. The acrylic will not crack under the rapid temperature change.

BENT ACRYLIC LAMPS

Many dramatic lamp designs can be made using a strip heater. The strip heater is, essentially, a heating element mounted in asbestos. Acrylic, which softens when exposed to heat, may be laid over these elements, heated, and bent in a straight line, at any angle. A popular and effective heating element is Briskeat RH-36. That is the element shown here; it is easily built into a frame of wood, aluminum foil, and asbestos paper by following the instructions on the package. It is inexpensive and readily available at hardware stores and plastics supply houses. To use the strip heater, prepare the acrylic just as you would to finish it: cut it to the proper dimensions, sand and scrape the edges, and polish where necessary. Of course, protective coverings must be removed. The acrylic may be marked with grease pencil along bend lines, but mark it on the side which will not be exposed to heat, and rub the pencil mark away later, when the acrylic is aligned.

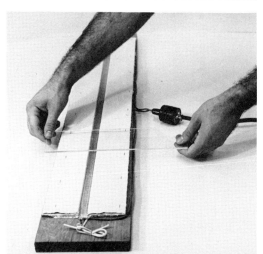

Center the mark over the heating element. If the acrylic is very long, prop up the ends so that the sheet lies flat on the frame. After aligning the acrylic, the pencil mark should be carefully rubbed off.

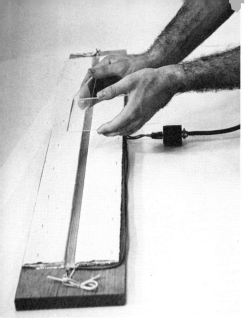
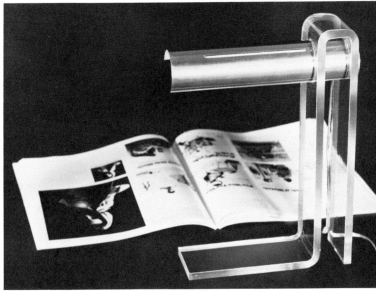

When the heated area begins to welt, it is ready for bending. The side which faced the heating element must become the *outside* of the bend. Fold the plastic to the proper angle, and hold it until it sets. Areas of plastic not directly over the element do not get hot.

This lamp, by Neal Small, requires only three 90° bends in the acrylic. With some planning you can drill the holes for the shade beforehand, or you can wait until the plastic has been bent to shape. Like other designs, the shade on this lamp rotates 360°. The lamp measures 12″ high, 9″ wide, and 3″ deep, and holds a 40 watt tubular bulb. *Courtesy: Neal Small Designs, Inc.*

This desk lamp by Neal Small is built of red, black, and yellow or white translucent sides which diffuse the light from a 40 watt bulb. The lamp involves three bends—two of which could be accomplished on a strip heater. The sweeping curve in the base line was executed by heating the entire piece of acrylic in an oven and then shaping it to the curve of a mold while the plastic was hot and flexible. *Courtesy: Neal Small Designs, Inc.*

The use of globelike forms is taken a step further in these designs—also by Neal Small. Rotationally molded polyethylene spheres, 16″ in diameter comprise this stack-up. *Courtesy: Neal Small Designs, Inc.*

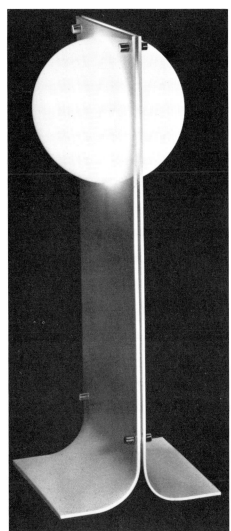

Neal Small employs the bending technique in two other effective designs. Each lamp is composed of two identical halves of white acrylic. The plastic globes are made by blow molding. The taller of the two is a floor lamp which stands 40" high, and is 14" wide. *Courtesy: Neal Small Designs, Inc.*

A veritable table lamp using the 16" spheres. *Courtesy: Neal Small Designs, Inc.*

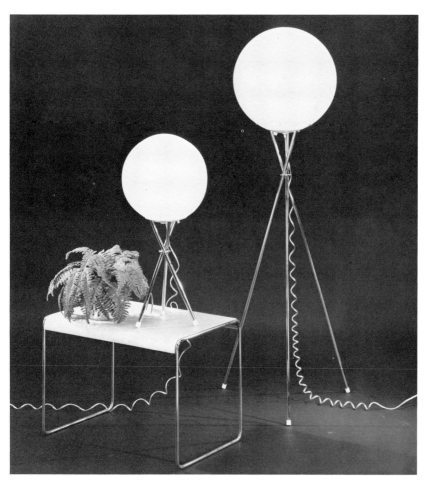

Although not as light as polyethylene globes, glass spheres are quite popular in lampmaking. These floor and table tripod lamps, for example, employ 12-inch- and 16-inch-diameter globes balanced on polished chrome tubes. The legs are attached to a circular rim which fits inside the globe and holds it in place. The table lamp stands 26″ high. The floor lamp is 58″ tall. *Courtesy: George Kovacs Lighting, Inc.*

A globe with a small lip is known as a *fitted glass shade*. This one rests upon a white plastic sewer pipe. Construction of this lamp is shown in chapter 6.

POLY-MOSAIC CYLINDER LAMP

Poly-Mosaic® is a versatile plastic craft material. The Mosaics are square plastic tiles which may be used as they are or fused in a home oven (as illustrated in chapter 7). They are available in a complete range of rich transparent and opaque colors. Poly-Mosaics can be readily glued to the walls of an acrylic tube, and the eighteen colors available increase the creative possibilities. To adhere the plastic tiles to flat surfaces, or to each other, styrene adhesives should be used. To adhere the tiles to curved surfaces like this cylinder, a silicone paste which comes in a tube, Silastic 732 RTV®, or clear silicone caulking, is used. A tile cutter, lower left, can be used to cut the tiles—as can a large nail clipper. Alcohol should be used to clean the plastic tube before beginning to work.

Clean the acrylic thoroughly with alcohol to remove all dirt and grease. With a wax pencil, sketch a design on the tube. Define the borders of the colors that will be used. These lines can be allowed to remain on the tube, since they will not interfere with the adhesion of the Silastic paste.

Apply the Silastic RTV to a small section of the tube which is ready to be covered with tiles. Spread it on with a palette knife or tongue depressors. It should be thick enough to completely fill the space between tile bottom and tube when the Poly-Mosaic is pressed firmly into the paste. Be certain to use adequate ventilation.

Press individual tiles into the silicone. Often, you will discover that at a certain angle whole tiles will fit smoothly along the tube's surface. But, in any event, tiles must also be cut to fill smaller spaces. The tiles should be placed as close together as possible to maximize color intensity. The inevitable lines between tiles can be filled later with acrylic modeling paste, which acts as a grout.

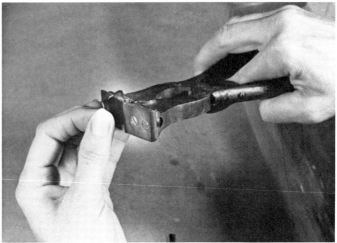

Poly-Mosaics can be cut into precise shapes with a tile cutter, or with large nail clippers.

Once the entire surface has been tiled, allow the adhesive to dry completely. During the tiling process, it is wise to remove any of the Silastic that oozes up between tiles. If some remains, cut the large lumps away with a sharp knife after it is dry. Acrylic modeling paste may then be smeared over the entire surface—to fill all cracks and spaces between tiles. The paste acts as a grout; it prevents light from leaking through spaces between tiles. Light leaks distract, and they detract from the intensity of tile color.

After the modeling paste has dried (it takes several hours), remove excess from the surface with a damp cloth.

147

The top and bottom edges can then be finished. File down the rough spots with a grinding bit placed in a hand drill.

Half tiles are then cemented onto the edge with styrene cement. Brush the cement onto the surface, and firmly place the tile.

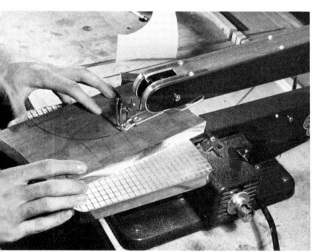

A base is constructed for the finished tube from identically shaped pieces of wood which are notched so that they fit securely together. A Dremel Moto-saw® cuts this 1-inch stock effectively.

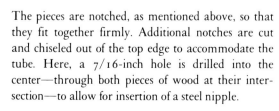

The pieces are notched, as mentioned above, so that they fit together firmly. Additional notches are cut and chiseled out of the top edge to accommodate the tube. Here, a 7/16-inch hole is drilled into the center—through both pieces of wood at their intersection—to allow for insertion of a steel nipple.

After drilling, the pieces are disassembled, sanded, and stained. This photograph illustrates most clearly the configuration of the notches.

The steel nipple, fastened with slip washer and nut at top and bottom, will support a socket, and conduct the lamp wire. The pipe also anchors the notched wood legs.

Illuminated by a tubular incandescent bulb, the brilliant colors of the Poly-Mosaic tiles shine through. Similar effects can also be achieved with glass, but, in this application, the colors, light weight, and ready workability of this plastic material certainly win out.

Black acrylic tubes, cut at an angle, weighted, and wired for a 50 watt reflector bulb allow Neal Small to effect an ingenious lighting solution. The light does not come directly from the bulb contained in the base. Rather, it reflects from a mirror on the inside of the tube's top. *Courtesy: Neal Small Designs, Inc.*

POTICHOMANIA LAMPS

Acrylic cylinders, wide neck glass vases, and hurricane shaped glass shades can be decorated with cut paper by a technique called *potichomania*. Package wrapping paper designs were precisely cut and arranged temporarily on the outside of this acrylic cylinder to test match and size the pattern. The paper will be glued to the inside of the cylinder. Any paper and any design can be used.

Very thin papers should be sprayed, on their back side, with a clear sealant. It stiffens them slightly, making handling easier.

The cylinder was cleaned with cold water detergent and water to remove dirt and retard static buildup. Mucilage mixed with a drop of glycerine acts as the adhesive. Acrylic emulsion, the kind used for painting, would also be effective. Small units of paper are glued on, one at a time. Apply pressure from the back, while looking through the front for air bubbles. Squeeze out all bubbles and wrinkles; if they remain they can cause blistering later.

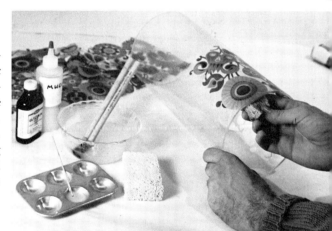

Once all the pieces have been glued into place on the inside of the cylinder, a coating of acrylic gesso is stippled over the entire inside surface with a sponge. This protects the design and creates a clear, but textured, coating.

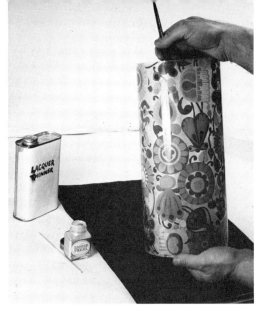 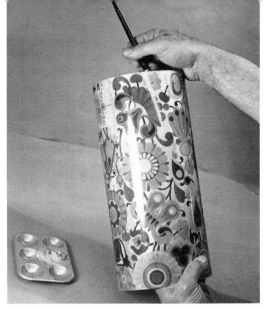

Two coats of metallic pearl translucent finish, which imparts a gold patina (Liquid Pearl®), may be painted over the entire inside surface of the cylinder as well. The first coat must be dry before the second is applied. Other metallic finishes, such as Treasure Jewels®, can be used too.

Metallic finishes take approximately one week to cure completely. When they do, a layer of acrylic paint can be added. Because the metallic pearl lacquer is translucent, the color of the paint will subtly enhance the print. Choose a color which will complement the design. Also consider the variations in texture that are possible. By streaking, or daubing or speckling the gold onto the cylinder, different effects can be achieved.

The completed cylinder sits on a round lacquered wood base, with a canopy cap covering the top opening. A natural linen shade adds to the warm effect of this table lamp.

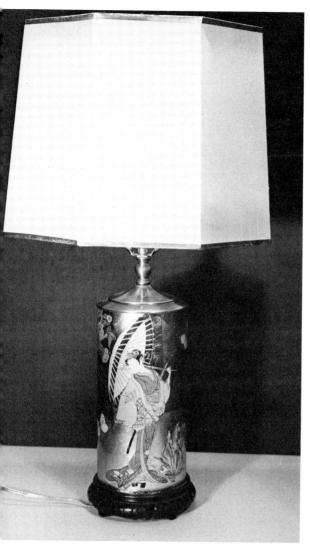

Gini Merrill combined a stylized Japanese print with a gold leaf background on this acrylic cylinder. She coordinates the lamp with an interesting hexagonal shade.

Glass can also be used for potichomania. This is a lamp in process. John Campbell and Lewis Morrow have cut and arranged flowers on a large vase in preparation for gluing.

Hurricane lampshades, like this one, are commonly available at lamp stores and many hardware stores. Dolores Greig prepares the glass by washing and drying it thoroughly. She then adheres the paper design with a white glue. The surface is then sealed with two thin coats of a clear acrylic spray, and five coats of acrylic gesso. New coats should be applied as the previous one dries.

John Campbell and Lewis Morrow finished this glass lamp base in the potichomania style by painting the inside surface of the glass with a white acrylic after attaching the paper cutouts.

MIXED MEDIA

In addition to using each material alone, a host of possibilities is opened to the lampmaker who can combine different media successfully. In the proper design and setting, any two materials will complement each other: wood and plastic, metal and stone, plastic and stone—and so on. The permutations are infinite.

GOOSENECK DESK LAMP WITH A MARBLE BASE

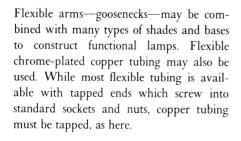

Flexible arms—goosenecks—may be combined with many types of shades and bases to construct functional lamps. Flexible chrome-plated copper tubing may also be used. While most flexible tubing is available with tapped ends which screw into standard sockets and nuts, copper tubing must be tapped, as here.

The flexible tube is passed through a hole at the back of a spherical metal shade. Notice the ventilation holes at the back of the shade. A porcelain socket screws onto the end of the tube, wired as discussed in chapter 2.

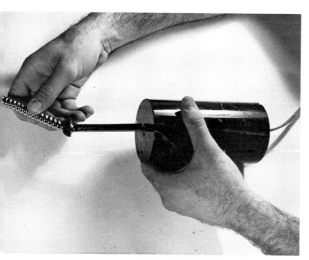

Predrilled marble bases may be obtained from foundries and dealers, although it may be necessary to chisel out a channel at the bottom of the base to conceal the lamp wire. The flexible arm is inserted into the marble. It may be secured with a nut or epoxy adhesive.

A reflector bulb completes this compact desk lamp.

An all chrome lamp may be constructed using parts available from a plumber's supply outlet. These sleek table and floor lamps were not. They stand 25" and 55" tall respectively. Beige butcher linen shades in a 1/2-inch pleat cover two 75 watt bulbs. *Courtesy: George Kovacs Lighting, Inc.*

This polished chrome floor lamp, 68" tall, provides 300 watts of indirect lighting. It has a dimmer switch on the back. *Courtesy: George Kovacs Lighting, Inc.*

The exciting combination of materials—chrome and granite—causes an otherwise commonplace table lamp to stand out. The overall height is 40", with a 16-inch linen shade. *Courtesy: Koch & Lowy, Inc.*

5

Ceiling and Wall Fixtures

Designs for ceiling and wall fixtures involve some of the same considerations as more portable lamps. The amount of light needed, decor of the room, technical elements, and budget all enter into the decision of which fixture to build and install. Technical aspects are dealt with under each design, whereas chapters 1, 2, and 3 cover more generalized yet pertinent considerations—like mounting hanging fixtures by junction box, socket wall bracket, or hook.

The purpose of this chapter—since it is impossible to legislate either design or design solutions for specific areas—is to illustrate some lighting alternatives, executed in a variety of materials.

As ever, the range of possibility remains enormous. Fixtures may be constructed from parts and kits designed for that purpose—or they may probe the limits of a material and find new applications. Nylon string, for example, becomes a weblike suspension around an acrylic core, and hemp twine may be transformed into a novel shade. Scored vinyl alone offers potential for exciting faceted and curved lamps in all shapes and sizes. The acrylics—here in sheet, rod, dome, and tube—again find application in light-catching, functional forms. Tubular knitted nylon jersey, used to make clothing, becomes a simple, stretchable lamp-skin.

These are a few of the alternatives. They suggest materials, techniques, and designs to inspire new avenues of exploration.

PLASTIC SPHERE

Many plastic parts are designed to be assembled into lamps. These molded plastic hemispheres, for example, were manufactured as lamp parts. The faceted surfaces effectively diffuse and transmit light from a globe-shaped bulb.

The hemispheres come with special clips which hold them together firmly but are, nevertheless, easily removable to allow for bulb replacement.

This lamp really sparkles.

STRUNG NYLON

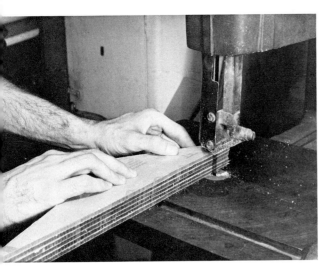

Lamps of acrylic strung with nylon draw their inspiration from the artist Naum Gabo. He experimented extensively with light and light-piping materials during the 1930s and 1940s. Though Gabo never constructed lighting fixtures, his creations are defined by light—which picks up the fine webbed patterns of nylon as light plays over the form. The concepts behind Gabo's forms are simple enough: fins of acrylic form a skeleton, and extruded nylon monofilament defines complex-curved surfaces between those fins. The nylon is held in place by fine notches cut into the edges of the acrylic. Here, notches consist of a single cut with a bandsaw. Different gauges of nylon, of course, will require notches of varying sizes.

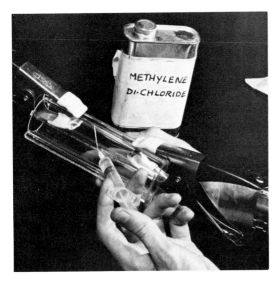

The fins are attached to a central core of acrylic tubing. Six square rods of acrylic are first adhered to the core with solvent cement. (See acrylic section of chapter 4.) The rods are held on with spring clamps while the cement is applied to the joint. Capillary action draws the solvent in, and the bond is secure in several minutes.

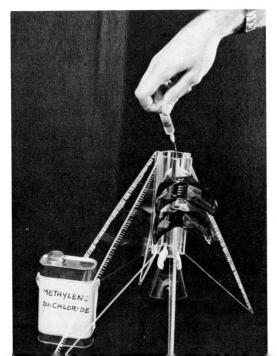

The notched fins are attached to the square rods in the same manner. They are held in place with spring clamps while the cement is applied and allowed to dry. The fins may be any shape at all; as you develop a sense of what patterns the nylon will make, your preferences for fin design will emerge.

Nylon cord is strung from top to bottom, bottom to top, etc. Begin by knotting the end—be certain that you have plenty on hand, too. Place that end in the topmost notch of one fin. The nylon should fit into the notch securely—if the nylon is too loose and is not tightly wedged, the notch is probably too large. Then draw the nylon down to the bottommost notch in an adjoining fin. (You may string rotating clockwise or counterclockwise.) Pull the nylon taut and slip it into that notch. The channel itself should play a large part in holding the cord in place. Draw the string to the topmost notch in the next fin—and so on and on, until nylon has been stretched between all the notches on the fins. Tie off the end with a knot just past the last notch to be filled.

A single cross of nylon creates a deeply scooped pattern on these sharply angled fins. Additional crosses of nylon, placed lower on the fins, could be added to achieve a more "filled-in" design. In fact, many lamps of this type do use several—or many—levels of very fine nylon.

"Plexima" employs two rings of acrylic as the core rather than the tubing. Eight acrylic fins create a denser nylon pattern. *Courtesy: Koch & Lowy, Inc.*

Fins of different shapes create significantly different designs. This Plexima, a Scandinavian design, illustrates the light-piping qualities of acrylic and nylon cord. *Courtesy: Koch & Lowy, Inc.*

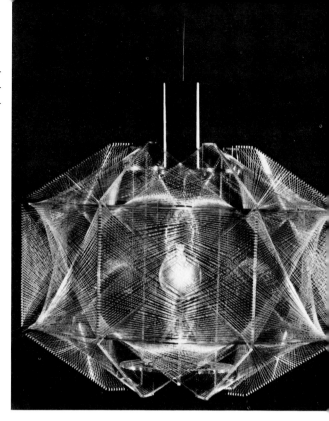

MULTI-UNIT ACRYLIC HANGING FIXTURE

Because acrylic is so easily formed and machined, this plastic has become a widely used lamp material. In this case, four pieces of acrylic were bent on a strip heater (see chapter 4) to form boxes that will contain glass globes.

The smoked acrylic boxes are pinned together at their corners using findings which screw together. The globe caps—which house the sockets and hold the glass in place with screws—are suspended by metal rods which fit into holes in the acrylic.

CEILING AND WALL FIXTURES 161

Glass globes are manufactured in many sizes, as are the fittings that secure them. Not only do they successfully diffuse the light, they also allow for easy bulb replacement.

This hanging fixture attaches to a ceiling box. Acrylic may be cleaned with spray cleaners for plastics or a mild soap solution, such as 10 percent liquid All ® and water. Always be careful, however, when using water, or even a wet sponge, near electrical connections. It is wise to defuse first.

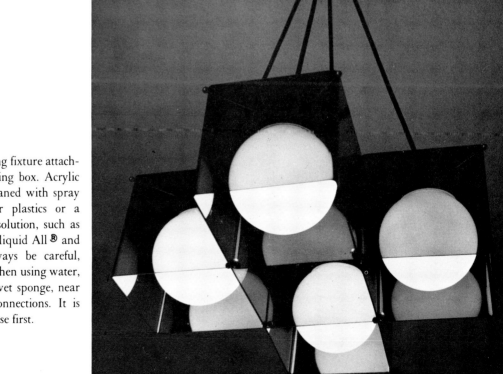

Acrylic may be formed into curves as well. To do so, heat the piece on an aluminum sheet in an oven set at 350° F. When the plastic becomes flexible, drape it over, or into, your mold. It will retain that shape when cool. *Courtesy: Mobilite*

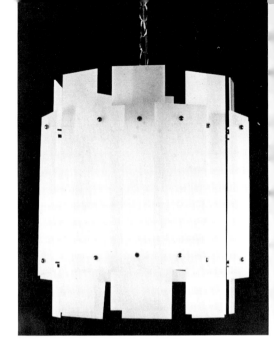

Acrylic is available in many colors and textures. The most common types are clear, smoke, black (opaque), white (translucent), white (opaque), and mirror, but most suppliers stock blue, red, green, purple, and yellow. Textured sheet suggests a design alternative in some applications as well. This fixture employs translucent white acrylic over a chrome frame. *Courtesy: Mobilite*

SCORED VINYL LAMPS
BY CURTIS STEPHENS

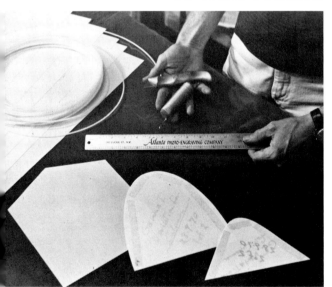

Curtis Stephens designs and executes magnificent lamps in flexible vinyl. He favors the translucent type, but silk-screened clear vinyl has application as well. Designers have also taken advantage of the flexibility of other types of plastic sheet including Mylar, acetate, thin acrylics, and fiberglass. Opaque white matte vinyl sheets are available from Firestone and Union Carbide; most plastics suppliers stock them. Make certain that the material is no thicker than .015 inches, however; anything thicker is difficult to work. The basic tools and materials include the vinyl sheet, nylon tubing 1/8-inch O.D., a sharp knife, a scoring tool with a rounded tip, and patterns for cutting and scoring.

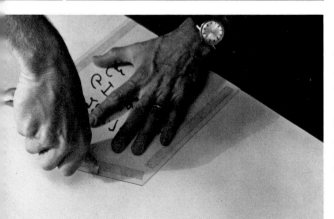

This technique lends itself to the use of modular units which are cut and scored from the same pattern. Curtis Stephens cuts lamp segments from 21" x 50" sheets with the aid of a pattern made from the same material. It is not necessary to actually cut the vinyl. Simply score it with a sharp or pointed tool, fold along the score line, and tear the vinyl.

The edges of each section must be scored to create a lip to be joined at the seam. To score, draw a dull point—comparable to the tip of a ballpoint pen—along the material, guided by pattern or ruler. Scores should be made on the outside of the fold.

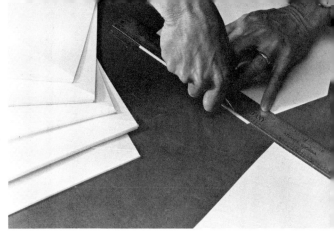

Curved folds should always be scored around a pattern. The larger and smaller curves will be folded in opposite directions and are scored alternately on opposite sides of the plastic.

Complete the scoring on each unit and then bend the plastic along these lines. The edges are first folded back for the seam; then, the curves are pressed and shaped. Curves, however, are flattened out so that the seams may be completed.

The folded edges are butted together and joined with 1/8-inch O.D. nylon tubing that has been slit down one side. Solvent cement—meant for vinyl—should also be applied along the seams to adhere the joints permanently. After the cement has dried, the curves are again pressed into shape.

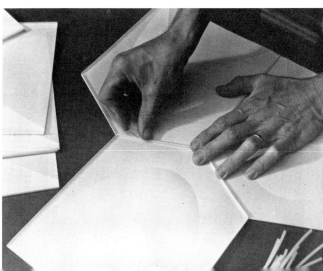

With all seams completed, curved folds are bent, and grommets are added at seam ends for additional strength and to allow ready attachment of lamp hardware.

As Curtis Stephens's work illustrates, large, graceful forms are possible with sheet plastics like vinyl.

Vinyl lamps may be constructed in any size—from the smallest shade to ceiling high units. The material will accommodate considerable complexity of design, though simple lines often work best. This series, *Courtesy: Curtis Stephens*

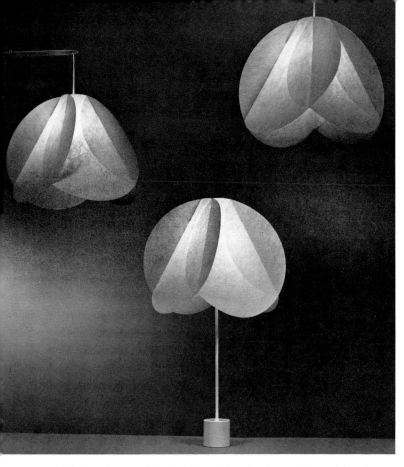

Circles of spun fiber laminated in plastic are slit and fitted together to form this wall/table/ceiling fixture. *Courtesy: George Kovacs Lighting, Inc.*

Wilhelm Vest of Austria designed this artichoke lamp in acrylic for George Kovacs Lighting. Because of the open spaces between the patterns of opal acrylic, heat dissipates easily and large bulbs may be used. *Courtesy: George Kovacs Lighting, Inc.*

Many pieces of flexible plastic sheet were combined in this form. Tabs at the sides of each unit fit into slits in the pieces below. Because the sheet is flexible, it will bend, accepting the curvature given it by this design. The tension and the tab-in-slot construction make this lamp structurally sound. *Courtesy: George Kovacs Lighting, Inc.*

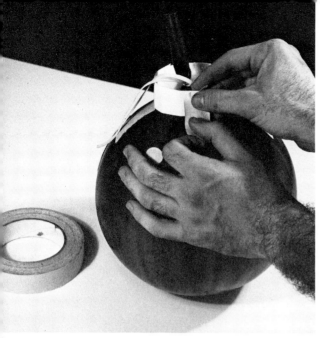

WRAPPED CORD OVER A DEFLATABLE ARMATURE

Balloons of rubber, plastic, or canvas serve as collapsible armatures for many materials. Often, papier-mâché is laid over inflated forms and, when it hardens, the temporary support is deflated and removed, leaving only a strong shell. That technique has been adapted here to the use of heavy wrapping twine. A rubber balloon is inflated to the desired size and tied off. Since a space must be retained to allow for placement and replacement of the bulb, a collar of cardboard is taped around the neck of the inflated armature. Shapes other than a sphere may be obtained by tying or taping around the balloon at different points.

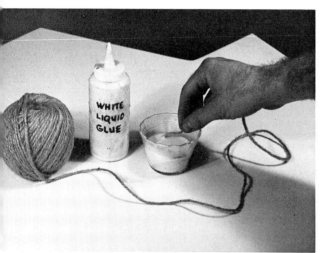

Any type of string or twine may be used to create this globe. The secret is to saturate the cord in a dilute mixture of white liquid glue. Add water to the glue and mix thoroughly until it has the consistency of whole milk. Then draw the twine through the solution. It should be thoroughly wet when applied to the form.

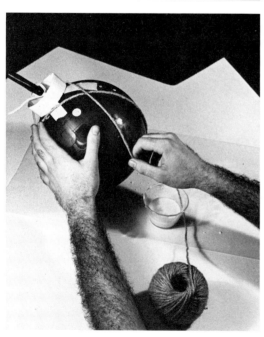

Wrap saturated cord around the balloon firmly. It should be taut, but not so tight that it cuts into the inflated surface. Cover the surface as completely as necessary to achieve a firm structure. The glue will stiffen and adhere the string when dry. The density of wrapping, beyond minimal concerns of strength, however, will depend upon individual tastes. The amount of light filtering through the form will be determined by how closely the twine is wrapped.

When the form has been covered, suspend it by the neck of the balloon to allow for even drying. When dry, deflate the balloon, remove it, and you will have a rigid shell.

The socket is joined to a wooden crossbar which will fit easily into the globe and support it.

The cord—up to the hanging fixture—is camouflaged by weaving it through a raffia chain.

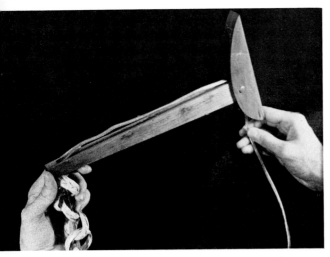

This type of device effectively suspends many hanging lamps from the wall. One feature is its adjustability; when lifted, the arm releases the wire, allowing the lamp to be raised or lowered.

Strung up in this setting, the string globe shines brightly.

The same hanging devices may be redesigned in many ways. Mobilite's version suspends a wicker basket shade on a walnut arm. *Courtesy: Mobilite*

Different types of baskets adapt very effectively to the requirements of hanging and ceiling fixtures. They are usually light, easily wired, and often naturally ventilated. This version was photographed in the Phillippines.

An Indonesian variation of the basket lamps suggests other alternatives. Chapter 6 further details the use of baskets in lighting fixtures.

WOOD STRIP LAMPS

The use of thin wood strips—slightly thicker than veneer—has long been popular in Scandinavian countries. Essentially, units of the thin wood are shaped (usually while wet), attached into units with a white glue, and several units are combined into a lamp. Several companies market kits complete with precut, preshaped, prenotched wood, corner joints, glue, and wiring. This kit, by Interdesign Craft, is one example. The cut, shaped, and notched wood strips are set into corner pieces and glued into place.

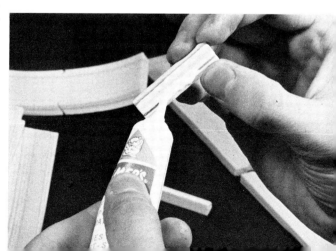

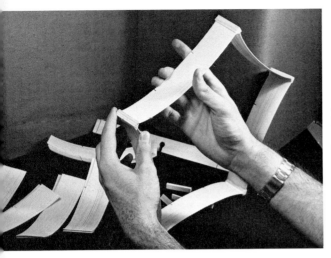

For this design, seven pentagonal units must be built. Each unit should be glued and allowed to dry fully before assembling the seven.

Notches at the top and bottom of each unit allow them to be securely stacked.

As they are added, glue must be applied to each point of contact.

The design is a clean one, and the fixture is quite attractive. While commercial units do give advice on the proper bulb wattage to be used, be careful when lighting one of your own invention.

STRETCH JERSEY TUBE LAMP

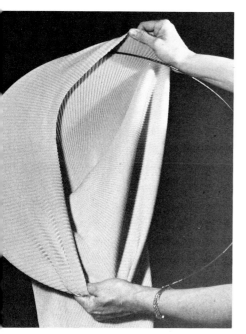

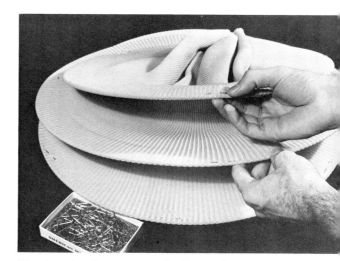

Pin rings in place temporarily. For variety, use different ring sizes.

Tubular stretch jersey is nothing short of magnificent as a lamp material. It is readily available at fabric stores, inexpensive, easy to sew—and it acts as a terrific light diffuser. Lampshade rings of different sizes are available in a complete range of standard sizes plus, of course, custom shapes. Insert the rings to the proper depths within the tube.

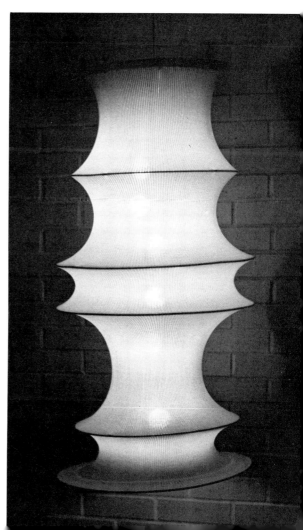

Tack rings into place, blind stitching them on the inside of the tube with nylon thread. The ends should be tucked under and hemmed. Remove all pins.

The top ring should be the same fitting as the top of any lampshade. The lamp will be hung from it as well. The wiring is quite simple: candelabra sockets attached to a single length of wire—and frosted spherical bulbs.

PAPER LANTERN LAMPS

Paper lanterns from Japan, China, and India are available from import emporiums in a variety of simple, pleasing shapes. They may be individually wired—or combined into lamps that reach from ceiling to floor or any point in between.

Paper lanterns may be joined with white liquid glue. Draw a bead around the form along the circle of contact and press the shapes together firmly. Allow each joint to dry completely before completing another.

Dramatic fixtures are possible. Plan out the order of shapes beforehand, and light the lantern with a string of Christmas bulbs, or make your own with a series of candelabra sockets.

MIRRORED CEILING DIFFUSER

Translucent acrylics enable lampmakers to create ceiling fixtures which emit very soft, diffuse light. This basic design employs an extruded aluminum frame—the kind available in most art supply, camera, and department stores. The frame contains a piece of acrylic mirror—with a hole in its center to attach the unit to the ceiling by threaded pipe. The frame pieces themselves are joined by steel L's which fit into grooves in the extrusion.

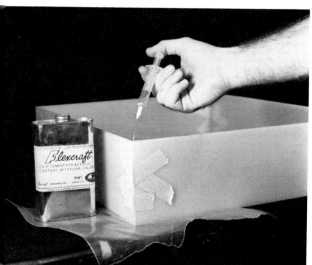

The band of translucent white acrylic, cemented at the corners . . .

. . . sits within the frame.

The completed ceiling fixture reflects the colors of the room.

Ceiling fixture in chrome; walls of mosaic glass mirror, bas-relief of mirrored acrylic.

ACRYLIC ROD FIXTURE

Acrylic rods were chosen to execute this nestlike ceiling fixture because of the unique way they respond to light. Whether the fixture itself is on or off, the rods are continually reflecting, refracting, and transmitting whatever light is available. The best cement to use with acrylic is PS-30®. It is a two-part system that dries clear and forms a permanent bond. According to the directions, the resin is measured out.

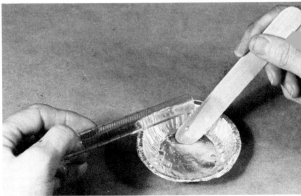

The two are combined carefully and mixed completely.

Then the catalyst must be measured.

Rods of two different diameters compose this nest. They were cleaned with alcohol and laid over a composition dome normally used for planting, just to impart the shape. Masking tape held the rods securely in place. The PS-30® mixture should be applied to every point of contact. The cement is extremely strong and a small daub at each connection should provide the strength necessary. Allow the form to cure overnight.

The sculpture covers a three-bulb canopy. Threaded pipe, covered with an acrylic tube, extends from the canopy's center, through the nest, where a circle of clear acrylic fits over the pipe, and is secured with a chrome end nut. It is the acrylic circle that actually supports the structure. Even when the electricity is off, the acrylic rods catch available light.

When the bulbs are on, the fixture sparkles. *Design copyrighted by the authors*

ACRYLIC DOME LAMP

Handsome lamps may be created by painting the inside surface of acrylic domes with opaque and translucent paints. Acrylic domes are available from manufacturers of electric signs, plastic fabricators, and skylight manufacturers. Begin by sketching the design on the outside of the dome with a grease pencil. Those lines may be rubbed away later. Since the dome is transparent, you will be able to see them while painting the inside surface.

To obtain different colors, it is necessary to mask the areas to be painted later. Liquid latex, or another masking material for acrylic, may be painted on the inside (concave) surface, following lines drawn outside. Test all masking materials on a small area first to be certain that they peel away easily and do not etch the acrylic.

When the mask has dried, spray on your first acrylic or lacquer color. Several coats will be necessary to cover fully. Hold the dome up to the light to determine how the surface will look. Apply thin coats and allow each coat to dry for a few minutes before applying the next. Paint applied too thickly will run. Always spray in a well-ventilated area.

Pull off the masking after the paint has dried fully. The remainder of the inner surface was sprayed with several coats of opaque white acrylic for two reasons. Not only does the light-dark contrast suit this type of lamp, but the white acts as an excellent diffuser.

A porcelain socket mounted on a plywood base provides for the illumination and support of the dome. Holes in the base allow for ventilation, but, nonetheless, this remains a small, enclosed area, and a low wattage bulb should be used.

The completed lamp may be table, wall, or ceiling mounted.

HANDS LAMP BY LEE S. NEWMAN

The most exciting lamps are often those that—while remaining functional—are completely zany. A case in point: The Hands Lamp. The idea is that hands hold everything, so why not have a set of hands hold a lighting fixture as well. Of course, *living* hands get tired and are difficult to attach to a wall without losing some of their strength, so molded hands substituted nicely. Jack Newman, our model, first coated his hands and forearms with Vaseline. This avoids painful sticking of hairs and nails later.

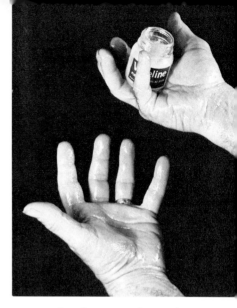

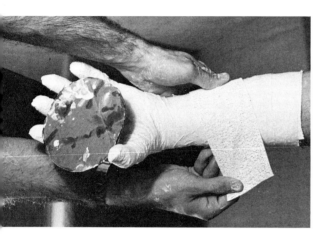

The hand and arm must then be wrapped securely with plaster bandages. This is the same material used to immobilize broken limbs while they mend. It is available at drugstores, from medical suppliers, and as a craft material (under the brand name Pariscraft®). To use it, dip the bandage in water and wrap it around the form, smoothing the surface as you go. Cut narrow strips of bandage with scissors to get between fingers and cover fine details. Our model held a piece of cardboard of the same diameter as the fixture in order to maintain the proper figure configuration. Use three layers of bandage wherever possible. By the way, the water used for dipping bandages should never be poured down the sink drain; that white powder is plaster of Paris. When hard, it creates quite a clog.

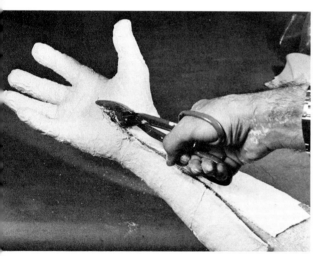

The bandage warms slightly as the plaster hardens—nothing to be alarmed about. When this happens, feel the shell. When it becomes hard enough to sustain the shape, it must be slit open and removed from the arm. Cut the bandage up the forearm with heavy scissors or wire nippers. Your model may be able to wiggle free without more cutting, but usually slits will need to be cut up to the bottom joints of a few fingers. Do not fret—your work has not been destroyed. Patch all cuts with additional pieces of bandage and add additional layers to sections which appear to be weak. Allow the shell to dry completely before carrying on the next phase.

Two arms will support the fixture; the other arm should be constructed in the same manner. Both will be filled with plaster of Paris. The right way to measure and mix plaster is illustrated here. Begin with a bucket—polyethylene is good because the plaster can be cracked away when it hardens. Fill it partially with water. Slowly pour the plaster into the center of the water so that a mound forms beneath the surface. Keep pouring, at the same point, until the top of the mound protrudes as shown here.

Plaster should be mixed by hand. Get right in there and squish it through your fingers until it is smooth and creamy.

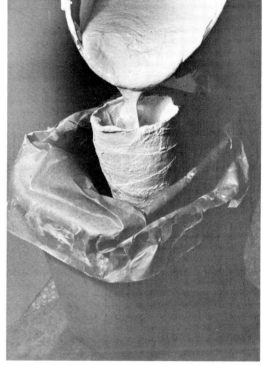

Be prepared to work quickly once the plaster is mixed since it hardens rapidly. One arm was propped up in a lined trash can. Pour the plaster in slowly at first so that it flows fully into all the fingers. One arm must carry wiring from the fixture to the base and the other must contain a bolt for attachment to the cylinder, so plan ahead for those necessities before filling. The plaster should set fully within an hour, but it will take several days for such a thick unit to dry out completely.

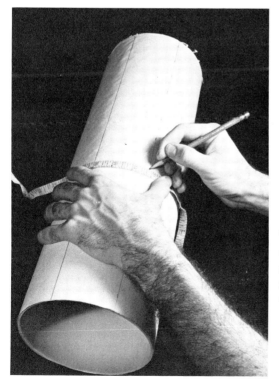

The plaster arms will support a fluorescent fixture encased in a mirrored acrylic tube. Since the tube is opaque, a section must be removed to provide illumination. A tape measure is used to measure the width of the planned opening in the cylinder.

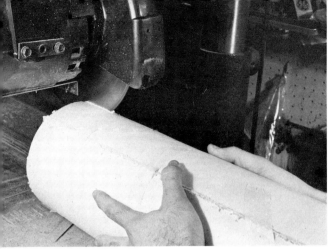

The cylinder is cut by ripping with a stationary radial arm saw. Dress the edges as you would any other acrylic—by scraping and sanding. Do not dress the ends, only scrape them to remove deep scratches, since they will be capped with circles of acrylic.

Before attaching end pieces and arms, mount the fluorescent fixture inside the cylinder. Most fluorescent fixtures come complete with transformer, starter, and end brackets. The enameled metal box is attached to the cylinder by bolts and end nuts set through holes in the surface. The black mark at the right is wax pencil, indicating the positioning. It rubs off. Coaxial cable, as illustrated later, extends from one end of the fixture through a hole in the end piece, into a hole in the thumb of one hand, down the arm, and into the base, where it connects to lamp wire and a push-button switch.

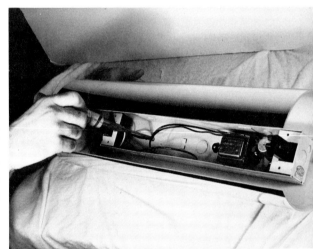

Clear acrylic circles cap the ends. Each requires one hole to accommodate the coaxial cable, or steel supporting bolt. PS-30®, an acrylic cement discussed in chapter 4 and used earlier in this chapter, is applied to the edge with a wooden applicator stick.

Set down the drilled circle—edges previously dressed and polished—carefully. Do not smear the adhesive.

Wipe away any excess cement that oozes out of the joint, and tape the circles down with masking tape to keep them in place while the PS-30® cures.

After the end pieces have been cemented securely in place, the arms must be attached. The arm shown here contains coaxial cable that will connect the fixture to the electricity source. The cable was threaded through a hole in the thumb. That hole was reinforced with a fast-hardening two-part epoxy to prevent chipping of the plaster and plaster bandage shell. The arm and fixture were then propped up so that the shell could be filled with plaster. A bolt was set into the palm of the other arm with plaster (*inside*) and epoxy (*outside*) reinforcements.

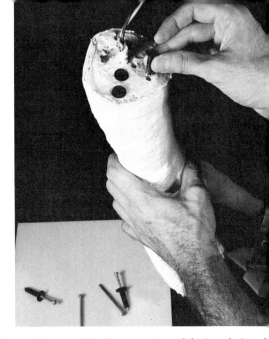

Excess length of plaster bandage is trimmed away with a sabre saw. It is easy—and recommended—to allow some excess which will be trimmed later. Try not to pour the plaster up farther than necessary, however, since hard plaster is difficult to cut and trim. Rather, add a little more plaster later, if absolutely necessary.

The plaster-filled arms may be attached to a base with any of the large variety of devices designed to be used with plaster walls and parts. The base, in this case, is wood, and the Hands Lamp will be mounted so that the unit will fit onto a wall. The plaster attachment device used here is known as a plaster wall plug. It consists of a bolt and a latex rubber sleeve. A hole is drilled into the plaster, and the sleeve is set into place. The arm is then set against the base, and a bolt is passed through the base, into the sleeve, and screwed into a set of threads at the end of the sleeve. As the bolt is screwed in, the threaded section is drawn toward the base, compressing the rubber sleeve. The tension against the walls of the plaster hole and the base created by the rubber sleeve keeps the fixture together. Five such units were used in each arm. More traditional attachment mechanisms may be employed as well. For example, bolts may be set head first into holes drilled into the plaster. After filling those holes with epoxy, the bolts will be permanently embedded, and the arms may be secured by passing the bolts through the base and fastening them with nuts. Always plan ahead, and, when drilling holes to accommodate bolts, wire, and switches, drill or chisel spaces to recess any findings which will appear on the back of the base.

This handsome fixture is innovative and provides excellent lighting for a desk or bed. A push-button switch mounted at the lower right allows easy access; a lamp wire runs from the back bottom edge to an outlet below the desk. *Design copyrighted by the authors*

6

Lamps from Found Objects

One of the most rewarding achievements in lampmaking is constructing a lamp from found objects—materials originally grown, made, or designed for other uses. The challenge here is in discovering an object's potential and redesigning another function for it. These may be natural materials such as gourds or driftwood, salvaged objects such as old wine bottles; or redefined things such as a coolie hat used for a lampshade.

Old shapes are given new meanings. They are collected and then recreated, usually resulting in a lamp that has great charm. Sometimes the original object and its use are identifiable; a wine bottle is obviously a wine bottle; other times origins are lost—such as using sewer pipe as a lamp base.

The range of materials is endless. A trip to a building supply supermarket, a "junk" store, your basement, or attic will yield a host of potential resources. These finds are combined into lamps no differently from materials designed particularly for lampmaking.

Some adaptations, however, may be necessary when utilizing forms in new ways. A wine bottle must be drilled to allow a neat exit for the wire. Metal objects may require soldering. Wood scraps need shaping and finishing. Gourds must be cleaned out. A weaver's shuttle is adapted with a wooden projection. Baskets are epoxied before drilling holes so the reeds don't unravel, and so on. Just a few possibilities are shown here.

184 LAMPS FROM FOUND OBJECTS

TECHNIQUES FOR RE-CREATING OBJECTS

Most conversions require attachment of parts just as in assembling any lamp. Usually, a hole is necessary at top and bottom of the object so that a threaded pipe—the backbone holding the wire—can be tied to a base, cap at the top of the opening, or whatever findings are used. Holes may have to be drilled. Sometimes parts can be epoxied together; other materials, such as wood, can be glued or nailed, the metals soldered, and so on. (Most of these techniques have been described throughout this book.) A few ideas and ways of mounting and transforming materials are described in this chapter, and more definitively in the accompanying photographs and captions.

Drilling a Hole in Glass or China

We would not recommend drilling a hole in a crystal decanter that is to be converted to a lamp base. It is too valuable and its worth will be destroyed. Ironically, drilling a hole in a discarded wine bottle will increase its value—when it becomes a lamp.

Drilling a hole is simple enough, but it is a slow process. Haste will cause the bottle or vase to fracture because a great deal of heat builds up through friction during the drilling process. Carborundum bits shaped like spades are available. These can be used in a portable electric drill. The only other necessary material is a small supply of kerosene. Kerosene lubricates the tool and the area being drilled, reducing some of the friction created while drilling glass or ceramic. The bit should be dipped into the kerosene frequently, and some kerosene should be sprinkled on the area being drilled.

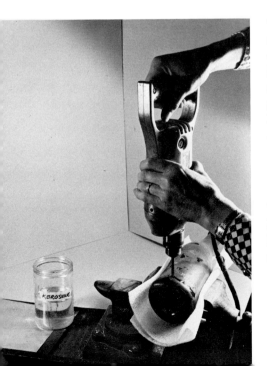

To drill a hole in a glass (or ceramic) bottle, securely mount the bottle in a vise and, using a carborundum bit for glass, proceed to drill. Every 20 seconds or so, allow the bit to cool off.

▶ The bit is dipped into kerosene periodically to keep the tool lubricated and sharp.

LAMPS FROM FOUND OBJECTS 185

After the hole has been made, the lamp wire is strung through the hole and fed through a steel nipple that in turn fits into a rubber plug. Here the plug is hammered into the bottle with a rubber mallet. After the plug has been inserted all the way to the brass cap that finishes off the "hole," the wire will have been mutilated by the pounding of the mallet. At this point, more wire is pulled through and the broken end is sacrificed.

A butterfly clip is screwed into a coolie hat (that was made in southeast Asia.)

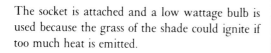

The socket is attached and a low wattage bulb is used because the grass of the shade could ignite if too much heat is emitted.

A tall glass vase that was converted into a lamp. It has a round walnut base and a hard fiberglass shade. The entire lamp is five feet tall.

It would be handy to have a vise on hand so that the object, protected from the jaws of the vise with cloth, can be held securely during drilling. It takes about a half hour to drill a 1/2-inch hole through the average wine bottle. Apply only moderate pressure, and pause frequently to allow the drill bit (and glass) to cool off.

Drilling a Hole through Woven Reed and Other Fibers

Most baskets, highly adaptable into lamps or shades, are made of reeds, vines, thin branches, or grasses of various types. Because they are woven and usually are "springy" in their woven state, cutting into a basket will trigger an unraveling process. To avoid this, a clear five-minute, two-part epoxy should be mixed and applied to both sides of the basket area to be drilled. After the epoxy hardens, a hole can be drilled through it just like any other hard material.

 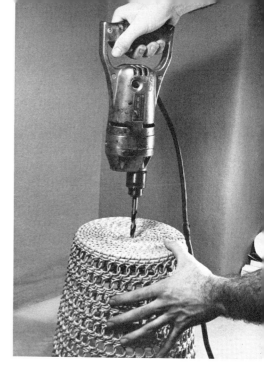

Before drilling a hole into a basket, a five minute two-part epoxy is mixed and applied to the area (top and bottom) that is to be drilled.

After the epoxy hardens, a hole is drilled with a 7/16-inch bit.

The wiring and pipe for hanging the glass shade or globe is inserted and the lamp is tied into the junction box in the ceiling. Glass globes, by the way, are available in a wide range of sizes.

The basket, now shade lamp, in its environment.

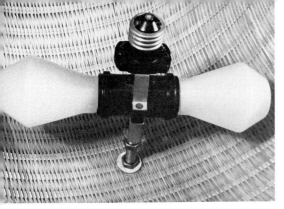

A twin socket adapter with built-on hickey for easy attachment to ceiling shades. This is a fruit and flower basket. Only low wattage bulbs should be used with potentially flammable shades like baskets.

The basket as a wall lamp.

The shade as a ceiling lamp.

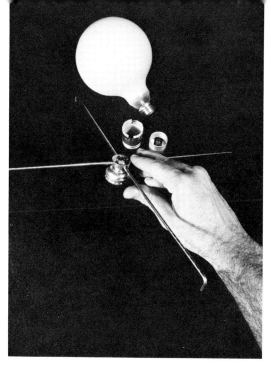

The completed hookup.

A spider attached with a slip washer under a hexagonal nut which in turn is screwed to a keyless socket. A large coated round bulb can be used as one alternative.

Another alternative is to use a glass shade or globe holder and fit it to the globe.

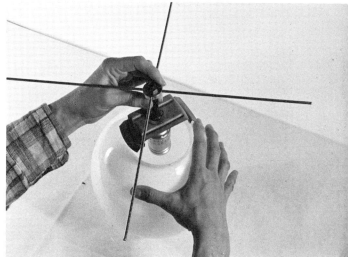

The spider hooks through openings in this basket and hangs from the ceiling.

189

190 LAMPS FROM FOUND OBJECTS

It is simple to drill a hole through a gourd. But it is not so simple to find dead center of an amorphous shape such as this. A level was used as an aid. Lamp pipe (1/8-IP) is the attaching device. A hexagonal nut screws on the steel pipe under the base, inside the gourd, and again on top of the cap. All of this holds all parts firmly in place.

CALABASH LAMP

A detachable harp is used with a standard base push-through socket, and a simple finial tops the whole.

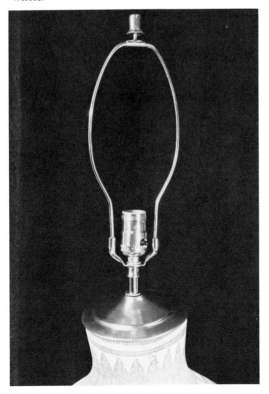

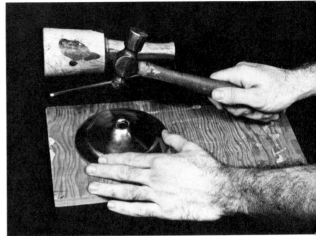

Because the gourd is so irregular in shape, a canopy converted as a cap for the opening is shaped with a ball peen hammer so that the pipe stands straight.

The gourd (or calabash) is from Oyo, Nigeria, West Africa.

Forming a Hole through a Gourd

Gourds are woody materials that take very little pressure to drill into. In fact, an awl or knife can be used to form a hole.

Connecting Attachments

Where hole making is impractical, adhesives are a solution, particularly the very strong epoxies. These two-part epoxies (we used a clear five-minute hardening type) do a great job of adhering dissimilar materials —such as glass to wood, metal to fibers, plastics to marble, and so on.

Weldwood®, a powdered plastic resin activated by water, is one of the best wood adhesives available. Titebond®, also mentioned in chapter 4, works well with wood, too.

White glues, which are polyvinyl acetate or chlorides, come in a variety of consistencies and are excellent for fabrics, papers, fibers, and some softer, thinner woods.

Besides adhesives, mechanical fasteners such as screws, nuts and bolts, rivets, and the like perform their special functions by connecting pieces that would be under too much pressure or tension for use of glues, or have too shiny a surface to be gripped well (under tension) by epoxy.

Where one alternative doesn't work, there always is another. Wire forms can be made to grip edges of objects much as a wire hanger snaps onto a plate used in wall displays. Wire can also be employed more decoratively to cage a piece such as a large crystal rock or a shell. Caging is actually wrapping of the form in a random manner so that the wire becomes an intrinsic part of the rock or shell structure. Caging would function well where a form would be destroyed or altered by permanent marring of its surface.

Similarly, antiques that become bases for lamps certainly should not be depreciated by a disfiguring attachment. One solution is to have a metal base or wooden base custom-made that fits tightly around the base of the form and connects with tubing (threaded at each end) from the base to the lamp socket and shade, somewhat like the way the wooden sculpture figures are attached in chapter 4.

The white tube is sewer pipe; the disks are cut of white acrylic and a hole drilled with a 7/16-inch drill. All the parts are here for making one of the easiest and most effective lamps in this book. White silicone is used as the adhesive.

One disk is adhered with RTV silicone at the base and another recessed at the top.

A hole is drilled into the side of the sewer pipe and the lamp is attached in the usual way—a hexagonal nut under the base screws onto the lamp pipe and one on the top is followed by a socket. An in-line on/off switch is attached to the wire.

A glass globe or shade fits neatly into the pipe. The proportion of this lamp design can change depending upon the height of the pipe and diameter of the globe.

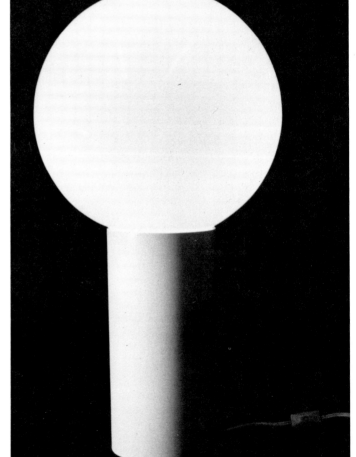

SOME FOND IDEAS FOR FOUND OBJECT LAMPS

Baskets, or various materials like tin cans, fiberboard containers (e.g., from ice cream stores), wire sieves, hanging plant containers, dome shapes, are a few found materials that could be converted, with some imagination, into lampshades.

Scrap wood, driftwood, toys such as blocks, dolls, and teddy bears, spindles used for stair rails and chair legs, tubes of various materials, apothecary jars, wooden boxes and small chests and other containers, old teakettles and antique kitchen implements, old tools, a stack of old but well-bound books, textile printing cylinders—are just a few ideas that could be used in transforming lamp bases.

Some multiple parts can be glued together and a hole drilled through the whole unit. Other softer materials can be encased in a clear acrylic box. Softer materials can be made permanent by coating the surface with a clear catalyzed polyester resin, clear urethane spray varnish, or even clear epoxy.

Keep flexible. Have fun!

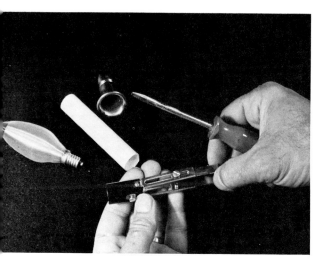

The height of an adjustable hickey is being regulated. A cardboard insulator goes over the hickey after it has been wired. The white candle sheath then slides over this.

The lamp is actually a retired shuttle converted now to a wall fixture to hold an electrified candle. A projection of wood was cut and drilled to fit over the spindle in the shuttle and project outward. A candelabra base socket on an adjustable hickey with a threaded end is screwed into a steel nipple which is tightened with a hexagonal nut underneath the sconce holder.

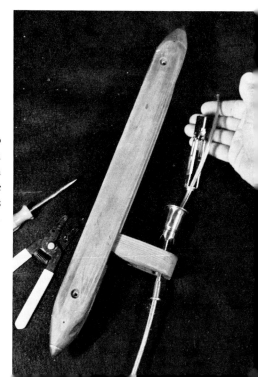

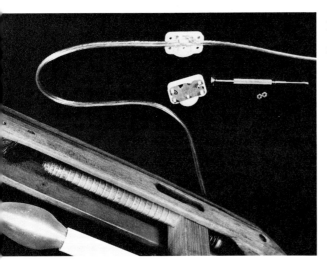

A clip on in-line high/low dimmer switch is used because part of the time this unit will act as a night-light.

The completed lamp with a flame-type bulb.

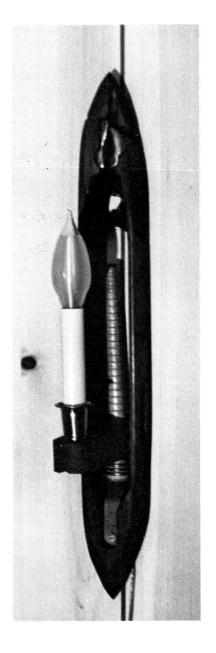

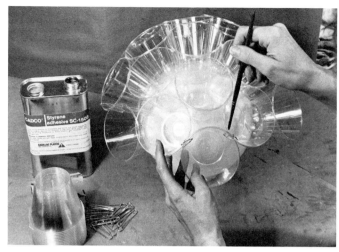

Disposable clear styrene cups are used to make a lamp. (Expanded styrene coffee cups are another alternative.) A Styrofoam® ball is the base mold over which the cups are placed. Paper clips hold them together and in place while a styrene solvent adhesive, SC1508, is run (capillary action) between the joints. When the solvent dries, the cups are adhered. Except now they are not cups but rather "crystal" elements in a chandelier.

Would you believe it!

7

Stained-Glass-Type Lamps

Glass has fascinated craftsmen for centuries. Beads and jewelry date from ancient times. There were plaster and alabaster slabs of the East, pierced and glazed in intricate designs; the cloisonné of Limoges with its narrow strips of metal separating colored enamels; and the custom of hanging tapestries over cathedral windows; but it took centuries for brilliantly colored transparent glass with lead caming to emerge from a craft to a high art. Windows were curtains of color. The effect they achieved is unique to stained glass—and one of the few modern materials which approach its intensity of color. Stained glass harnessed the most powerful and dependable light source known to man—the sun. The results have been spectacular. The light that filters through into a darkened or candlelit area has a spiritual, thrilling quality. On a sunny day the colors jump out. In haze, stained glass creates a diffuse, ethereal light. The secret is color intensity. Stained glass encompasses a complete range—from the palest water whites to fully saturated blue and red. Artists know that intense colors, subtly juxtaposed, can have enormous emotional impact—that is the effect that stained glass can have.

BACKGROUND

The beginnings of the stained glass craft are lost in the blur of the distant past. We know that the windows of Saint Sophia at Constantino-

Modern stained glass lamp in an intriguing shape, by Jack Cushen. *Courtesy: Jack Cushen*

ple were glazed, but whether or not the original glass was colored is questionable. The technique of stained and leaded glass, as we know it, was probably not used until several centuries after the Advent of Christ.

Legend accounts for the discovery of gold-colored stain. The story is that a young Dominican brother, who was working in the craft, was about to place some painted pieces of glass in the kiln when he was called from his work. As he rushed away, a silver button loosened from his clothes and dropped on one of the pieces. When the glass was fired, a lovely yellow stain was found where the button had rested. This craftsman later became Blessed James of Ulm and is honored as the patron of glass craftsmen. Indeed, white and a bright yellow were the first colors of stained glass windows.

Some bits of stained glass history emerge up to the eleventh century, but we know that the richly colored windows of that era grew from the less colorful ones in the past. The good Abbot Suger of Saint Denis played an important part in shaping stained glass interpretation in the twelfth century. The twelfth and thirteenth centuries yielded some of the most magnificent windows of all ages. The cathedral at Chartres is a treasure house of twelfth- and thirteenth-century jeweled windows in full rich color. Those traditions continued to the nineteenth century when craftsmen of the Louis Tiffany school learned how to make translucent glass and thereby eliminated the costly and time-consuming job of painting and firing each piece of glass. (That was necessary to control the diffusion of color around the edges of the shapes which was otherwise lost to the eye when intense light struck the glass.)

Stained glass lamps are part of a much more recent tradition. The materials are the same, but the interpretation is somewhat different: to utilize the same intense colors in order to soften and shield light, to decorate, and to produce a warm glow. The original form—a dome-shaped shade suspended on a pedestal or from the ceiling—was popularized by Louis Tiffany during the early 1900s. Though Tiffany did not invent the style, his elegant, fanciful designs generated enormous interest in the possibilities of stained glass in lighting. The inevitable commercial copies abounded, even then. Tiffany's contemporaries contributed new ideas as well. Art nouveau, the genre of that age, flourished—and the stained glass lamp was one novel and exciting part of the forms and styles that emerged, an attractive new solution for the need to shield the electric light bulb and diffuse its glaring light.

Stained glass masterworks—windows, lamps, boxes, mirrors—from many eras have survived. European cathedrals still boast monumental windows hundreds of years old. What these forms reveal is a consistency of goal, a consistency of technique, and—until recently—a consistency in design. The goal of working with stained glass has remained very much the same since the craft was first mastered: to display as much glass color as possible with a minimum of connecting framework. It makes perfect sense, after all, to attempt to maximize color and minimize the interstitial elements that support the form. Techniques have remained quite consistent as well. Lead caming, discussed below, has always been used to construct stained glass forms. Today, of course, other valid and effective techniques have also emerged. These are labor-saving solutions for today. Such changes are understandable. Before the recent upsurge of interest in Tiffany-style shades, the glass of every era bore its own stylistic benchmarks. Today, however, too many people are more concerned with copying an earlier design, with mimicking another's thoughts and interpretations of the medium. We think it is much more satisfying to create your own designs, even though the process and materials have ancient origins, and to work it in a traditional manner. For this reason, no patterns are included here. Experiment creatively.

MAKING THE PATTERN

Stained glass lamps begin with a small-scale drawing, which is then transposed into a full-sized master pattern called the *cartoon*. Using carbon paper, the cartoon is traced onto a sheet of brown wrapping paper. That brown paper pattern is then cut out along the outlines to create templates for the individual pieces of glass which must be cut.

The scissors used by stained glass craftsmen have a special feature: a third blade which cuts a thin sliver of paper off the edge of the pattern when the brown paper copy is being cut apart. That sliver provides space for the heart of the lead caming that will run between the pieces. Otherwise, it is necessary to compensate by trimming a small amount of paper from each template. This is almost unnecessary when adhesive-backed copper tapes are employed, since the copper is very thin and the pieces may usually be tapped into place without difficulty.

Once the template has been cut into individual units, arrange the paper pieces juxtaposed on a flat surface. Traditionally, glassworkers attach each piece to a glass easel with a beeswax and Venetian turpentine compound (a modern substance like double-stick tape or Plasti-Tak® works well). The pieces are then numbered to indicate what color glass will fill that space. Then all of one color is cut—perhaps blue—then all of another color and so on.

The actual cutting is done around individual templates. The brown paper should be attached to the glass temporarily with double-stick tape or Plasti-Tak®. Then cut directly around that pattern (see glass cutting below). Once cut, the glass replaces the paper pattern in the layout. When all the pieces have been cut, the lamp or panel is assembled.

Patternmaking for lamps is somewhat more specialized than patternmaking for flat panels since the glass must conform to a three-dimensional surface. Some master craftsmen—like Jack Cushen—make their pattern as they go along, shortcutting the traditional procedure. Dome-shaped lamps are usually constructed over a mold. Traditional molds are of some hardwood, such as maple, and they are available—or can be made—in an infinite variety of shapes. Modern molds are often adapted parts—like plastic hemispheres of the proper size. The piece-by-piece pattern is constructed by laying a piece of brown paper over the section to be constructed and filling in lines for the pieces to be cut. Then cut apart that pattern, cut the glass, and fit it into place before going on to the next section. That sort of construction technique requires a good mental plan of your lamp. But it can also be adapted by making a sketch—which you transfer and fit into place section by section.

What you will realize as soon as you begin to piece the glass units together is that, depending upon the curvature of the mold, the pieces of glass cannot be too large. If they are, they will not fit around the form closely enough to approximate the curve. Other stained glass techniques do involve shaping individual sections of glass, but this goes beyond the scope of this chapter.

There are more precise methods of making a pattern for glass shades as well. The primary method is to fit paper over the form. Since most molds are rounded, it will be necessary to slice the paper at various points, draw it together, and tack it into place. Once the paper follows the curved contours of the form, sketch the lines onto it that will represent individual pieces of glass. Remove it from the form, flatten it out, and make a brown paper template pattern. It would be a good idea to save that original fitted pattern for use on later lamps.

In the case of panel lamps—or any lamp that is constructed of several flat sections soldered together—the original patternmaking process will suit admirably because the attendant complications of fitting to a curved mold are avoided.

TYPES OF GLASS

Stained glass useful to lamp craftsmen is available in two forms—transparent and translucent. Within those two categories, however, an enormous range exists. Not only are there infinite varieties in color, but there are variations in grain, thickness, intensity, and purity as well. Every type serves its purpose in a particular application. Craftsmen generally plan not only a design, but also for qualities of the glass which will be used to execute the design. When selecting glass, observe the changes in coloration throughout the piece—and, when cutting specific forms, check to make certain that the color intensity combines well with the color intensity of bordering pieces. Observe variations in the grain—will ups and downs look choppy or do they lend excitement?

Stained glass is available from specialists in every large city. Many are highly specialized—carrying the glass of only one factory or only one country. Use the telephone directory to locate sources. Another source is broken and discarded stained glass panels and lamps. If a piece has been damaged or discarded, there is nothing wrong with using the glass again.

CUTTING GLASS

Glass cutting is an essential skill. It requires practice, and patience, but by following the proper procedures, anyone should be able to master it after a short while.

The first step is to clean the glass. Use an ammonia-based liquid which will leave the glass free of grease. Then, lay the glass on a flat surface—wood, cardboard, or Neolite® is best. Place the brown paper pattern on the glass, holding it with double-sided masking tape or Plasti-Tak®, and cut directly around the pattern.

Glass cutters are available at most crafts and hardware stores. There are different types—but the primary difference is in the cutter wheel. Some are harder than others and will last longer—they also cost more. If held and used properly, all should cut glass precisely. Another variation is the ball end of the glass cutter. Some cutters have a heavy metal ball on the grip end which many craftsmen find useful in tapping along under the score line to initiate running the cut and aiding the break.

Grip the glass cutter, as illustrated, between the index and middle fingers, and use your thumb to support the cutter from the underside. Hold the tool firmly, but do not tense your arm muscle. Place the cutter edge down on the glass until you hear the sound of the wheel cutting into the glass, then begin to draw the cutter along the pattern line—all the while applying just enough pressure to hear the wheel cutting into the surface. Applying the proper amount of pressure takes some practice. It's a good idea to experiment upon a few pieces of inexpensive or scrap glass. A properly scored cut should break readily—that is the test. A ragged line could produce a cut that will run off in an unwanted direction.

Break each cut as it is made and never attempt to go over a score line: that damages the blade. Breaking the glass can be accomplished in several ways. To break a large piece of glass, lay it with score line parallel to the edge of a table—and slightly over it—and snap it apart. But if you are breaking long strips or small pieces, pliers may provide the best solution. *Running pliers* are wide pliers especially made for breaking long strips of glass; usually the larger part of the glass is gripped in one hand and the part to be broken away is gripped with the pliers. For smaller pieces, the same method may be employed using *grozing pliers* in one hand—or two pairs of grozing pliers. As before, experiment by cutting different shapes, different lines, and different curves. You will soon get the "feel" of how to score and how to break glass.

JACK CUSHEN'S STAINED GLASS LAMPS

Jack Cushen is a master glass craftsman who has adapted the best of the traditional and contemporary in glassworking techniques. To make this Tiffany-type stained glass lamp he employs an acrylic hemisphere. A hole is cut out at the top to accommodate the 1/2-inch brass ring which serves as structural support at the top of the form. First, size the brass ring; then cut it and solder it. The ring should fit snugly into the opening.

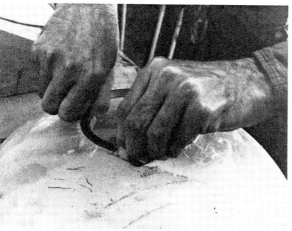

The next step is to tack the ring in place by attaching four pieces of glass. Those pieces serve to keep the ring level while other elements are attached. Since the lamp will, ultimately, hang from this ring, it must be level, and those elements that butt against it should be firmly soldered. To make certain that the curvature will be correct, Jack Cushen marks the glass against the curve.

Although glass cutters will eventually get dull and worn out, with simple care most will remain serviceable through many applications. Occasionally dipping the cutter into kerosene before cutting will extend the life of a cutter. Store unused cutters in a small jar with flannel or steel wool at its bottom and add enough kerosene, or kerosene and a light oil, to keep the wheel and axle covered.

The glass is cut along that line in one clean motion. Notice how the glass cutter is held: this is the proper form.

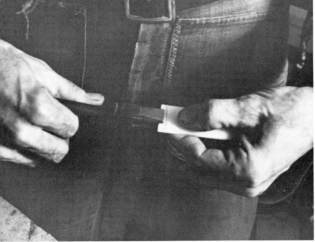

Small pieces of glass are broken away with grozing pliers or "breakers." Accurate cutting is important. The more finely the pieces fit together, the more fine the soldered lines will be, and the stronger the final piece.

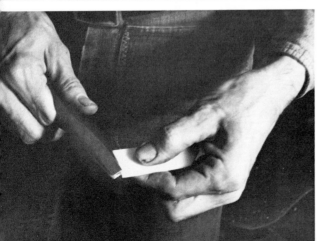

Rough edges are smoothed with grozing pliers. The edge is, essentially, nibbled clean.

Foil is wrapped around the perimeter of each piece of glass. The object is to cover top and bottom with equal amounts of copper tape. Different widths of tape are available. The type being used here is Scotch Brand copper electrical tape. It is recommended because its adhesive works well and does not disintegrate during soldering.

Once wrapped, the tape is snipped off at the proper length—there should be very little overlap. The tape should be pressed down onto the faces of the glass, and the corners tucked in.

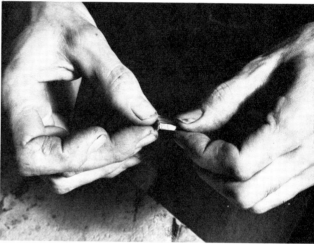

The foil is burnished onto the faces—but not the edges—with a metal or wood burnishing tool.

Jack Cushen likes to cut away excess copper foil from the face of the glass, and to shape it somewhat. It gives him better control of the solder, since solder will adhere only to the copper.

Four pieces, as mentioned above, are tacked to the brass ring in order to keep it level. Before tacking—which is a soldering operation in which just enough to keep the form in place is applied—both ring and copper foil must be coated with flux. Fluxes remove the oxides from metals and allow solder to effect a solid bond. A good soft solder flux is a zinc chloride based liquid.

A little bit of solder is applied to the iron to "tin" it, but the basic operation requires that the parts to be joined be heated and that the solder be applied to them. Because they are hot, the solder melts and fills the joint. The best solder for this type of work is 60/40—never use flux core solder, it makes a mess on the foil. Jack Cushen uses a 125 watt iron, or a soldering gun for tacking pieces together.

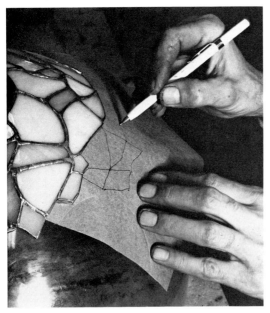

He then removes the paper from the mold, numbers each piece so that he can reassemble the order easily, and cuts along the lines.

Once the four leveling pieces have been tacked into place, he begins to make patterns for small sections. Although some craftsmen like to plan out the entire design well in advance, Jack Cushen begins with the basic idea in his mind, and, with that, plans out smaller sections one at a time, designing spatial patterns as he goes along. To do this, he slips a piece of brown paper beneath the last area completed, outlines that edge, and sketches in the new pieces.

He now has individual patterns for each piece. And each pattern is used directly. He places the paper over a slightly larger piece of glass and cuts around.

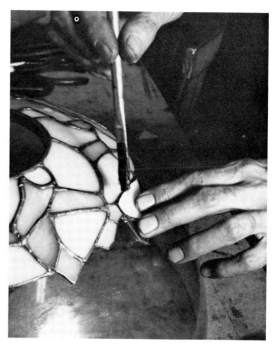

Once all of the individual units from a section have been cut, wrapped with foil, and burnished, he adds them to the lamp. Flux must, of course, be applied to the foil before soldering.

Jack Cushen likes to tack all of the pieces in place by soldering them at a few points, before finally soldering over the entire joint.

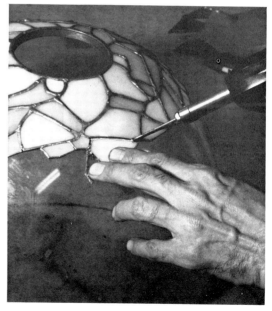

Once a section is in place, he adds more solder, but the outside of the lamp is not truly soldered over until it is finished. Before doing the outside, Cushen recommends that the inside be completely soldered, because the heat from soldering the inside will affect the outside. The main strength of the stained glass lamp lies in proficiency of soldering—an average lamp will require 2 to 3 pounds of solder. Solder can also be finished or patinaed in many ways.

With this tour de force—a magnificent stained glass dome executed completely in circles—Jack Cushen illustrates the use of a wedge (at the center). By maneuvering this device and by turning the dome, almost any part of the form can be made level, which facilitates flowing on of solder.

A finished stained glass lamp by a master craftsman, Jack Cushen.

207

For this lamp, glass panels were heated until flexible and shaped over a mold.

Stained glass lamp and planter—"plantern". *Courtesy: Glassroots Studios, New York. Photo by Nathan Rabin*

FUSIBLE PLASTICS

Prior to modern times, glass was the only material that could offer transparency and translucency and combine brilliant colors with either of those two qualities. Today, plastics allow an alternative. Fusible thermoplastics—like Poly-Mosaic®—are available in a wide range of colors. Each material has distinct qualities and benefits, and each should be recognized and used for what it can offer.

Plastics provide an alternative to glass which has never existed before. The Poly-Mosaic® tiles illustrated here are readily fusible in a home oven or broiler oven, come in a wide range of colors, can be texturized in a variety of ways, and are a fraction of the weight of glass. In addition, after fusing they can be cut to shape with a band or jigsaw and sanded by machine or by hand.

The processes for using this material are straightforward. Once you have drawn a pattern, cut out the paper shapes. On an aluminum cookie sheet, arrange the tiles generally in the shape of that contour. The tiles may be cut then with tile cutters or large nail clippers, or the tiles may be fused first and then cut on a band saw or a jigsaw. Place the cookie sheet in an oven or broiler oven at 350° F until the material reaches the desired degree of fusion (several minutes).

The tiles may be fused only slightly, removed from the oven, and overlaid with tiles of a different color or the same color and then replaced in the oven. This will result in color change and different textures, depending upon the amount of "cooking" and the number of layers and combinations of tiles. The longer tiles are "cooked," the flatter and more fused the final piece will be. It is wise, however, at least to fuse the Poly-Mosaics® until the edges weld together so that a strong piece will be formed.

After fusing, the tiles may be allowed to cool naturally; they will pop away from the aluminum sheet. Then the shape may be finely trimmed to the shape of the paper pattern on a band saw or jigsaw.

From that point on, the procedures are precisely the same as they are for cut pieces of stained glass. The Poly-Mosaics® may be constructed with lead caming or with adhesive-backed copper tape. Both methods are illustrated below.

CONSTRUCTION AND ATTACHMENT TECHNIQUES

Two alternative attachment techniques are presented here. One is the traditional lead caming (in combination with nontraditional plastic); the other is nontraditional adhesive-backed copper tape (in combination with traditional glass). Both are valid techniques. Lead caming is still used extensively in larger stained glass forms but, increasingly, skilled craftsmen are employing adhesive-backed copper tapes on smaller forms. The reasons are clear. Although lead provides great strength, the copper tape is a faster, easier material to work with and it is lighter—a benefit on smaller forms.

The technique of using thin channels of lead to hold intensely colored pieces of cut glass in place offers a perfect wedding of materials. Malleable, easily attachable lead fits readily around precisely cut pieces of glass. Though the lead is soft, the rigid material it encases allows craftsmen to create large, structurally sound units.

Lead caming, as these channels are called, comes in two configurations called H channel and U channel because the cross section of each is shaped like the letter. H channel is used between two pieces of glass with an edge fitting into the channel on each side. U channel—with only one groove—is used to finish edges, usually on the outside of the form.

In the traditional working techniques shown below, U channel caming is laid along a row of nails which define the perimeter of the panel. Each piece of glass is individually cut and placed in position and cames are soldered into place to contain each unit, while nails hammered into the wood below hold the pieces tightly in place. Wherever possible, unbroken pieces of lead caming are continued throughout the panel, since continuous lengths create a stronger form than many short strips.

The preferred solder is know as 60/40, which stands for 60 percent tin and 40 percent lead. (Make certain that you use solid core, wire solder.) Some solders contain a flux core, which only creates ugly oxides on the caming. Fluxes are necessary to prepare the surface of the lead to accept the solder, but most craftsmen find that a liquid flux—the brands recommended for specific solders are available—works best. It should be brushed on immediately before soldering. Its function is to remove oxides that prevent chemical bonding of solder and metal.

ADHESIVE COPPER TAPE

A modern stained glass technique employs adhesive-backed copper electrical tape. (Scotch Brand tapes are highly recommended because the adhesive does not burn off when heat is applied during soldering.) One attribute of the copper tape is that it may be shaped after it has been applied to the glass. The copper tape is wrapped around the edges of each glass unit and is burnished flat. Each unit is soldered to its neighbor over a plastic, metal, or wooden mold. Soldering proceeds in precisely the same manner as with lead caming. The copper is first painted with a flux, and then solder is applied—first to "tack" the pieces together at a few points, and then to coat the tape entirely. The solder outline can be controlled by trimming the copper tape with a sharp knife before soldering.

The use of copper tapes is relatively new. It employs a material originally created for use by electricians in making circuits, but this ingenious application has been pioneered by some of the finest contemporary glass craftsmen. Although this alternative to traditional lead caming is not as strong, for most smaller applications, like lamps, it provides more than enough support. It has the additional advantage of permitting greater detail, and it allows the use of smaller pieces.

A MOLDED POLY-MOSAIC LAMP

Poly-Mosaics® are square plastic tiles which are readily fused in a home oven. Because they are so easily fused, molded to take on the shape of any form, so readily cut, and available in a broad range of transparent colors, this material is ideal for lamps of many kinds. (The same material is used in a cylindrical lamp in chapter 4.) To fuse this plastic, place the tiles on an aluminum cookie sheet, and put the cookie sheet into a home oven or broiler oven set at 325° F. The tiles may also be cut to size with tile clippers or nail nippers.

After 2 minutes, remove the fused tiles from the oven, and press them firmly together with an aluminum spatula, to make certain that they fuse together fully and there are no gaps between pieces. Once the tiles have fused together, usually in about 3 to 5 minutes, remove the aluminum sheet from the oven. Slide an aluminum spatula under the melted tiles to separate them from the sheet and . . .

... press the fused pieces into a mold. In this case, we employed an aluminum colander, because it had precisely the domed shape we wanted. The Poly-Mosaics will cool in a few minutes and retain the curvature of the colander. Although this material is nontoxic, normal precautions should be taken to provide for good ventilation in any working areas. Also, never leave any working material in an oven unattended.

Poly-Mosaics will retain the molded shape after cooling.

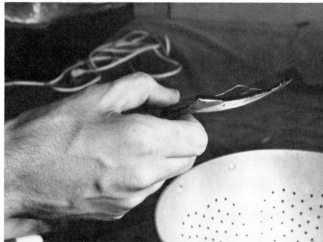

The material may then be cut with tile cutters, a band or jigsaw. The Dremel jigsaw works very effectively.

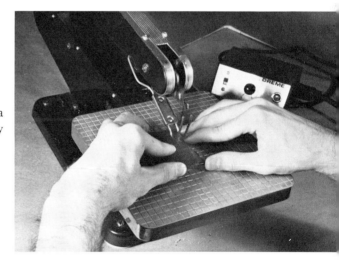

As pieces are completed and cut to the precise size—just as in making the stained glass lamp—place them beside each other inside the mold.

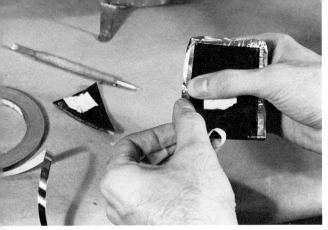

The plastic pieces are wrapped with copper tape just as glass is. The plastic, however, is much lighter.

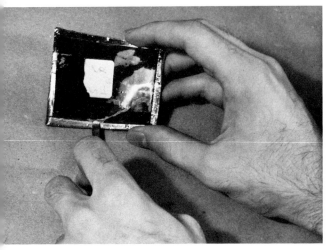

As illustrated above, burnish down the copper electrical tape, and fold the corners down.

Begin tacking the pieces into place. Since this mold allows us to work inside of the dome first, flow solder over all the joints on the inside after tacking the form together.

The dome, with the inside soldered completely, is removed from the mold easily.

Because Poly-Mosaics melt under heat, the outside can be readily "polished" by glancing a propane torch flame over the surface.

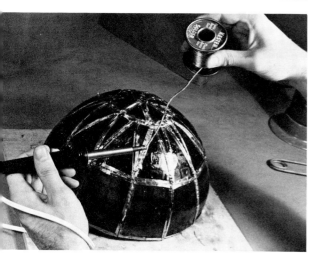

Apply flux and solder to the copper foil.

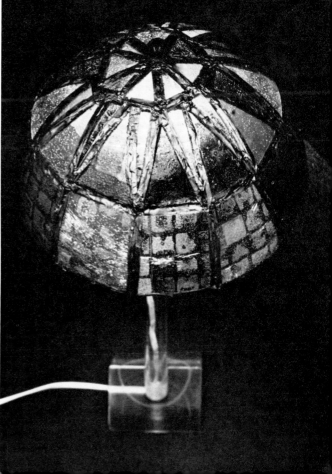

The finished lamp is a strong and handsome structure. The lines and textures are one of the intriguing facets of this material. Still other textures can be created during the heating process.

TRADITIONAL STAINED GLASS TECHNIQUES IN PLASTIC

This cylindrical lamp is constructed of units of Poly-Mosaic tiles. Each unit consists of six tiles, arranged in a rectangle and fused together in a home oven, as illustrated above. The construction technique which is employed is the very one developed by master glass craftsmen centuries ago. After nails are driven into the wood along a starting line, H channel stained glass caming is laid against the edge. Units can then be fitted into the channel and be temporarily braced in place by nails.

Strips of H channel caming are cut and soldered into place between each unit of fused Poly-Mosaics. The channels, soldered into place around each unit, complete a grid—with the pieces tightly encased within.

Having completed the flat panel, the form should be carefully bent into a cylindrical shape. Breaks in the cames created by stress can be mended by resoldering. Any gaps between plastic and the caming should be filled with steel-filled epoxy putty.

This type of lamp can be constructed in almost any size. The fact that the plastic is so much lighter than glass allows for a larger structure.

8

The Lamp as Sculpture

Light brightens. It enables us to see, but it often functions as a medium quite removed from the concerns of even and efficient illumination. Light has a sculptural quality. Light—directed, diffused, contained—can modulate surfaces. It can create a mood. Bright lights foster activity. Dim light relaxes. Variegated light intrigues.

Light is a medium in itself. Fine artists use light to highlight and illuminate their works in glass and plastic. Others employ amplified light beams to create three-dimensional works that float in space. For still other contemporary artists, neon and fluorescent lights become elemental sculptural units. Combinations of lenses and filters have allowed artists like Julio Le Parc to create graphic masterworks of light beams focused on mirrored, undulating, and moving surfaces.

Light fixtures can be works of art as well. Where designers, architects, and artists have turned their energies to lighting design, striking lighting solutions have resulted. Lighting fixtures are, after all, three-dimensional forms—and there is no reason at all why the implicit functionality of lighting should inhibit the creation of significant and practical forms. The category, however, is an elusive one, and it suits no purpose to attempt to define it precisely. Sculptural lighting fixtures do have characteristics in a sense. They are objects in their own right. A sculptural lamp might embody a unique material—or it might have a dramatic and exciting structure reflecting both technical and design insights. The

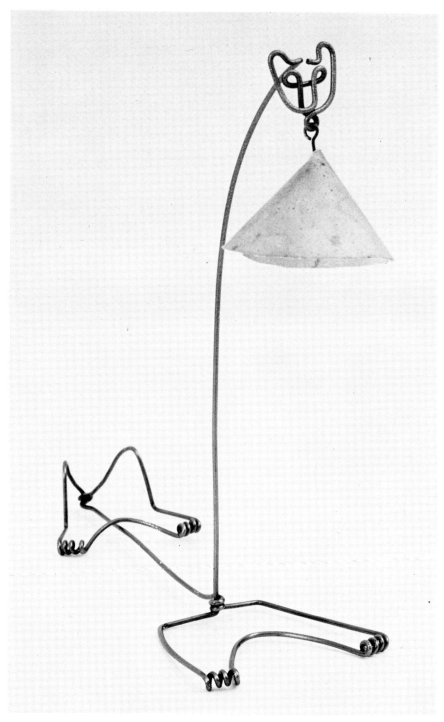

Sculptural lighting fixtures are not limited to the practical by any means—as Alexander Calder's nonfunctional *Cat Lamp* illustrates. The armature is iron wire, the lampshade, paper. (8 3/4" x 10 1/8" x 3 1/8".) *Courtesy: The Museum of Modern Art*

218 THE LAMP AS SCULPTURE

quality of light the lamp gives off—the mood it creates—becomes part of the form as well. Often, lamps with sculptural qualities provide soft, "mood" lighting—more ornamental than essential. But elegant lampworks can be, and are, designed for every lighting application.

The most valuable asset in striking upon innovative and exciting solutions to any problem is an open mind. Be on the lookout for new techniques, new materials, and new possibilities offered by the circumstances of a particular room or area. With that beginning, the range of flexibility and possibility is endless.

Illuminated polyethylene globes compose lighting fixtures on a variable scale. This one approaches the monumental—but a single unit (or more units) will create a different fixture. *Courtesy: Neal Small Designs, Inc.*

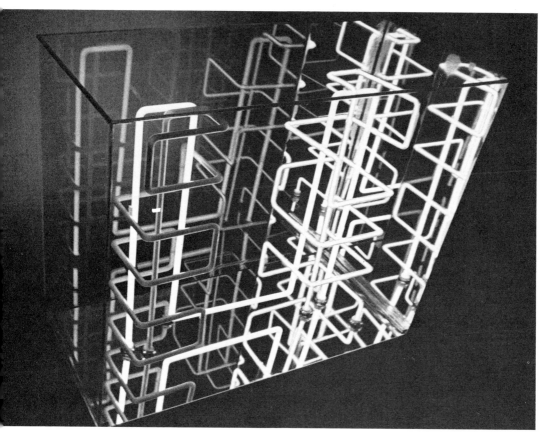

Neon, formed in three colors and patterns, multiplies as reflections in the side and back mirrors of this acrylic sculpture by Thelma R. Newman. A transformer is housed in the black acrylic base along with an interval switch that plays each color/light in alternating rhythms.

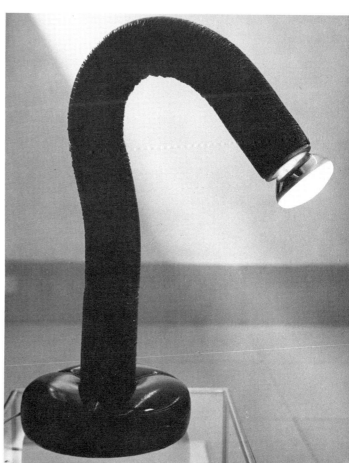

A reflector bulb screws into a housing made of a flexible gooseneck stem that is covered with foam and an outer skin of vinyl. This soft lamp, from Italy, with its balanced base has the capacity to be curved into countless sculptural shapes.

Wendell Castle employs globe-shaped light bulbs to punctuate his sculptures of fiberglass and polyester resin.

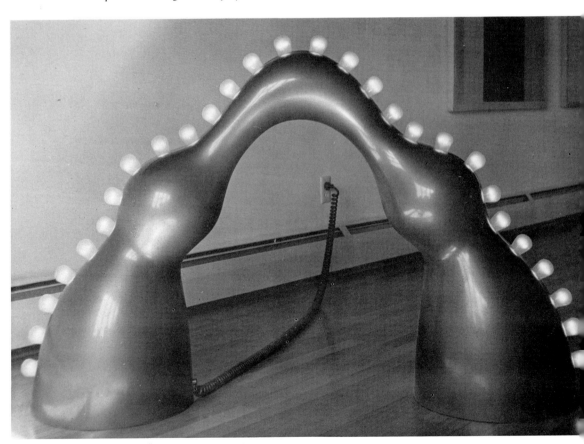

LIGHT BOXES

Light boxes are invaluable in many applications. They provide the perfect solution for slide viewing, more than that, they dramatically illuminate transparent and translucent materials. Colored and clear glasses and plastics are highlighted most effectively this way. Boxes may allow direct piping of light through an opening of controlled size, or larger lighted surfaces may be created using translucent diffusion panels—usually of acrylic or a frosted glass. There is no reason why a light box cannot be expanded to table size.

The light box really is a simple structure—often no more than a wood, metal, or plastic box with fluorescent or incandescent bulbs mounted inside. The choice of fluorescent or incandescent will depend upon the application (the room, the purpose of the box, the colors desired, etc.), and, since the bulb will often be contained in a small boxed area, heat becomes an important consideration. Fluorescents generate less heat per unit of light than do incandescent bulbs. For this reason alone the former should be used whenever possible. In any event, all light boxes require ventilation of some sort; incandescent light boxes must be ventilated carefully—with air holes at the base, bottom, or whatever. And every light box should be tested, by leaving it on under observable conditions for an excessive amount of time, to evaluate its heat buildup safety.

The light box is an invaluable device for underlighting *any* object, but it is especially effective when transparent or translucent forms are being displayed. Wood, metal, or plastic may be used to fabricate the basic structure. In this case a pine box was covered with black Formica®. The bottom of the box is lined with aluminum foil so that the light reflects up, toward the translucent plastic panel which diffuses the light. A pencil-thin fluorescent fixture is attached to one side. This smaller size fluorescent was used because there was not enough space to accommodate a larger bulb of any type—and, in any case, it provided enough light for our purposes.

Most often it will be convenient to place a switch in the wall of the box, for easy access. Toggle, push button, or rotary switches may all be used effectively. A box like this one, meant to supply light to a larger area, requires a diffusion plate. The most effective material we have found is translucent white acrylic. It is light, readily available, easily cut to size, and eminently practical, since it can be cleaned with a mild soapy solution.

The finished light box being used to display a sculpture in clear acrylic by Thelma R. Newman. Transparent forms really require the accentuation that effective lighting provides.

THE LAMP AS SCULPTURE 223

Zodiac, (6" x 10"), by Phillip Borden is an etched acrylic panel, highlighted by light from the box beneath it. When the surface of acrylic is abraded in any way, light transmits or leaks through the cuts, becoming a glowing line.

Jane Bearman's acrylic etching, *Reclining Figure* (9" x 14") is set into a light box as well, piping light through the etched lines.

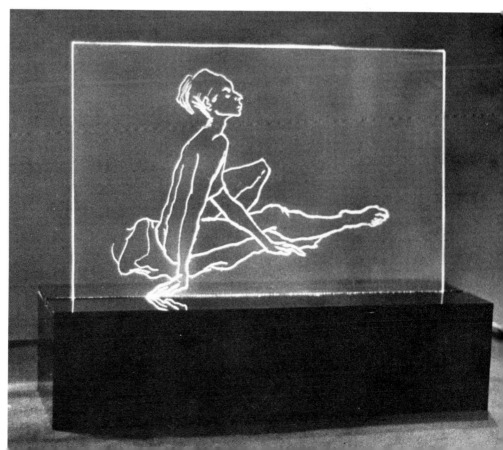

Thelma R. Newman's light sculpture *Light Sphere* is constructed entirely of clear acrylic domes and rods. It sits in a cylindrical light box of stainless steel. The light source is a single incandescent bulb; holes around the base provide ventilation.

Gae Aulenti's lamp consisting of acrylic fins on an aluminum base utilizes the light box and acrylic's property of piping light. It projects radiating light patterns. *Courtesy: Gae Aulenti*

Ecapuzzo's *Undula* is a sculptured light box. Opalescent and translucent crystal diffuse the light. *Courtesy: Koch & Lowy, Inc.*

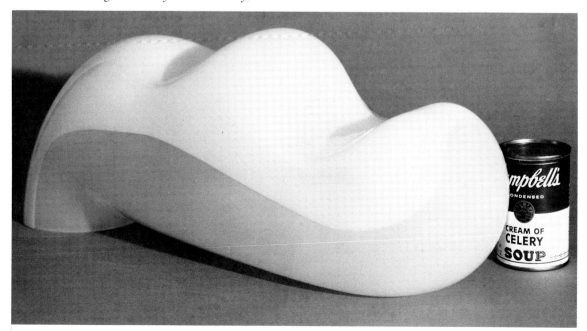

Acrylic rods pipe light up to a transparent globe in this light sculpture by Thelma R. Newman. The light box which contains a hole at the top is stainless steel.

Diffuse light is emitted by various configurations of bubbly crystal in this series of sculptured constructions designed for Koch & Lowy. *Courtesy: Koch & Lowy, Inc.*

Ugo Lapietra's lamp incorporates three hemispheres of clear acrylic. Each is drilled to create domelike holes that catch the light and articulate the curves of the hemispheres so beautifully. *Courtesy: Ugo Lapietra*

MIRROR LAMP

Mirrored plastic is a marvelous lampmaking material because it is easy to work with and its reflective properties provide a wide range of design ideas and alternatives. This mirror sculpture employs a 12-inch incandescent bulb to create a very soft, dramatic light. The panels of mirror both reflect light and contain it. The base is an acrylic-clad plywood box.

MIRROR LAMP SCULPTURE

A plywood box is cut, sanded, and assembled. A hole is drilled in the center to accommodate a bakelite socket housing. After wiring the socket, the unit is screwed to the underside of the wood box.

The box will be clad with a layer of black acrylic. Grooves in the acrylic will accommodate sheets of mirrored acrylic which will stand perpendicular to the base. The box is painted with a black, alcohol-based stain to prevent any white wood from showing through.

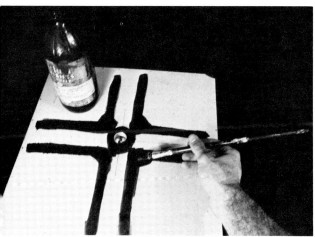

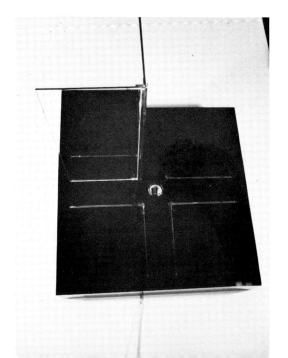

The sheet of black acrylic must be cut 1/4 inch larger than the top of the wooden box, to allow for the thickness of the side pieces of acrylic (1/8 inch thick). The acrylic top is routed, and then glued to the wooden box with two-part, five-minute epoxy. Side pieces of black acrylic butt against the top and are glued to the wood box. Eight pieces of mirrored acrylic (1/8 inch thick), with cut and polished edges, fit into the channels routed out of the black plastic base. The mirror is supported by aluminum trim normally used in building countertops with Formica. Again, with five-minute epoxy, the mirrored pieces are adhered to the aluminum trim and to the wooden base through grooves in the black acrylic.

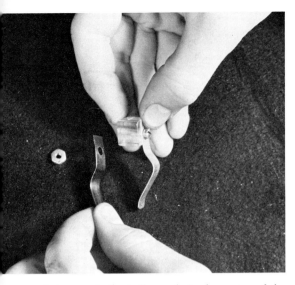

A long, tubular bulb stands in the center of the lamp, as tall as the mirrored acrylic fins. To eliminate glare from the front of the bulb, steel spring clamps are bolted to a small acrylic block.

The acrylic block, in turn, is glued to the back of a circular piece of mirrored acrylic. This unit clips to the front of the bulb.

The finished lamp by Lee Newman produces unusual reflections of the light which parallel the stark black, blue, and silver acrylics. (The mirrored acrylic comes with a blue backing.)

PIPE LAMP WITH SMOKY BULBS

This lamp employs various lengths of aluminum pipe—the type used for shower curtain rods and other lightweight supporting structures. The pipe is cut and then glued together with a fast-hardening epoxy. The result is simply stunning. With a wooden base—to avoid the starkness of unaccompanied aluminum—and smoked bulbs in candelabrum sockets, an elegant and adaptable lighting solution emerges. The same structure is easily adapted to wall fixtures, ceiling units, and even hanging lamps of various sizes.

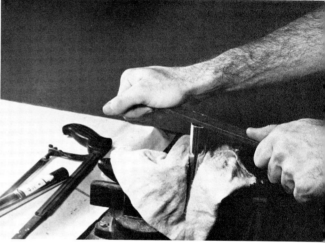

File away rough and uneven edges using a flat file. In some applications it will be important that the plane of the pipe end be perpendicular to the pipe.

Aluminum pipe is available in many diameters and lengths. The size chosen for this fixture accommodates a candelabra base socket which has been wrapped in electrical insulating tape. A smaller diameter tubing has also been employed for design effect. Always cut metal pipe with the proper tool and blade. Most often, a hacksaw (as shown here) will serve best. Use a smooth, back and forth motion, holding the saw blade angled almost parallel to the floor. Cut with the tubing in a vise whenever possible. This makes cutting easier and more accurate. Always wrap the tubing to prevent scratching, and tighten the vise jaws carefully to avoid collapsing the tube.

Remove burrs from the inside edge with a rattail (round) file.

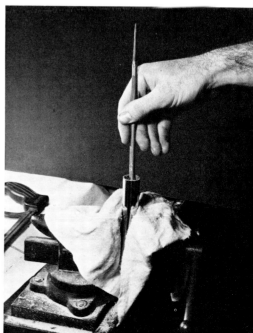

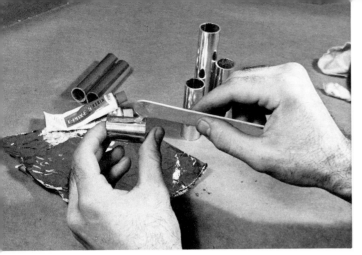

Two-part epoxy cements the aluminum pipe parts together adequately. First clean the metal to remove dirt and grease. Then, mix the glue. Regular or five-minute epoxy will serve well. Of course, the faster-curing type eliminates some working problems and saves time. But be careful —it hardens quickly. Spread the mixed resin along the line of connection with a wooden tongue depressor or a mixing stick.

Press the glue-laden pipe sections firmly together. If using a fast-curing glue, you may be able to hold the pieces until they set. Otherwise, tape the sections to hold them in place, or prop them up in a quiet spot.

Standard-sized mounting brackets are available for standard diameters of metal pipe. Here, a space is being chiseled into a piece of wood for one such bracket.

Holes were then drilled at the proper spots to accommodate wiring, and the bracket is being screwed into place.

The lights—there are five—should be wired in series. The main circuit—leading to the plug—is cleared of insulation to allow for attachment of leads from each socket.

The wood, having been sanded smooth, and the edges chamfered (angled with a file), is painted with linseed oil. This oil brings out the natural color of the wood and protects it from excessive drying and cracking. Apply two or three coats of oil, as it is absorbed; let it seep in overnight, and then wipe away the excess.

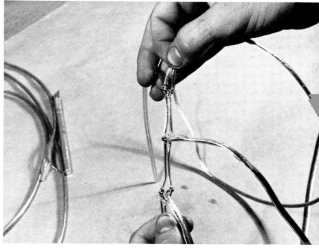

One lead attaches to one line, the other to the other. Each splice is performed as shown in chapter 2 and then soldered.

The splices could also be wrapped with electrical tape. The wiring was then set into channels chiseled out of the underside of the base. Keep the wires in place with plastic tape. Be certain, however, that you do not oil the bottom—the tape will not stick to any oily surface.

A good means of insulating the exposed sections is to coat each splice with a thin layer of a silicone sealant. This paste will harden overnight to create a permanent, watertight insulation.

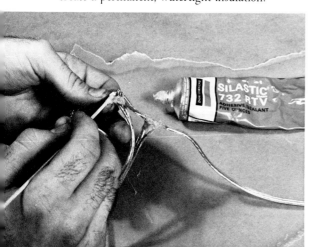

234 THE LAMP AS SCULPTURE

Wires to each light are passed up through the respective pipes, and the total pipe form is pressed down firmly onto its anchor bracket.

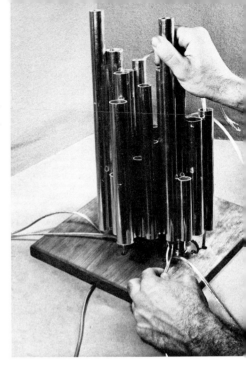

Candelabra sockets are wired as usual and wrapped with one layer of electrical tape. In this case, the single layer of tape provided a snug fit into each tube. If the tube is larger, use a thicker tape wrapping. Do not use a smaller tube, however, since safety requires at least *some* separation of socket from metal.

Rubber washers are set into the pipes containing sockets to provide a finished look, and to protect the bulb.

THE LAMP AS SCULPTURE 235

The finished lamp is a dramatic, steppe-like sculpture.

Supply Sources

BASES

Gim Metal Products
164 Glen Cove Rd.
Carle Place, N.Y. 11514
Cast metal bases

Little York Products
70 Hudson St.
Hoboken, N.J. 07030
Wooden bases

Marb-Velle
589 Ferry St.
Newark, N.J. 07105
Marble bases

BULBS

Duro-Lite Lamps, Inc.
17-10 Willow St.
Fair Lawn, N.J. 07410
Decorative and standard bulbs

General Electric Co.
Nela Park
Cleveland, Ohio 44112
Complete line of bulbs

Lite-Iron
P.O. Box 20796
Dallas, Texas 75220
Decorator bulbs

ELECTRICAL COMPONENTS

Leviton Manufacturing Company, Inc.
236 Greenpoint Ave.
Brooklyn, N.Y. 11222
Complete line of electrical components

Lighting and Lamp Parts, Inc.
West Babylon, N.Y. 11704
Packaged lamp parts

COMPONENTS—NONELECTRICAL

(chains, globes, harps, swivels, tubing, etc.)

Aetna Pipe Products Co. of Illinois
3515 W. Armitage Ave.
Chicago, Ill. 60647
Pipe and tubing—threaded, beaded, flared, custom bending, and spacers

Allen-Stevens Conduit Fittings Corp.
29 Park Ave.
Manhasset, N.Y. 11030
Loops, finials, loop hangers, spacers, swivels, ceiling hooks

Cable Electric Products, Inc.
Providence, R.I. 02907
Lamp kits, lampmaking products

Chilo Manufacturing and Plating Co.
2106-10 South Kedzie Ave.
Chicago, Ill. 60623
Harps, finials, loops, necks, risers, washers, shade fittings, glass shade holders, seating rings, rods, etc.

Craftsman Wood Service Co.
2727 South Mary St.
Chicago, Ill. 60608
Switches, brass tubing, harps, bottle converters, sockets, shade bullets, pipe, bulb clips, vase caps, finials, shade risers, couplings, bases, chain

DeRosa Lamparts
514 N. La Brea Ave.
Los Angeles, Ca. 90036
Lamp and lighting fixture parts—a complete line

Freeman Products
86 State Highway 4
Englewood, N.J. 07631
Vase caps, candle cups, ferrules, bases, pinup brackets, necks, check rings, ring loops, brass balls, finials, nozzles, locknuts, switch knobs, bushings, nipples, crossbars, hickeys, chain, decorative hooks

Highwall Metal Spinning and Stamping Co., Inc.
633-643 Berriman St.
Brooklyn, N.Y. 11208
Balls, bases, brackets, reflectors

I.W. Industries, Inc.
Hexagonal nuts, chain holders, decorative hooks, loops, lock washers, spacers, tubular arms, stems, nipples

Kirks Lane Lamp Parts Co.
1445 Ford Road
Box 519
Cornwells Heights, Pa. 19020
Sockets, chimney holders, harps, fittings, pipe

Minnesota Woodworkers Supply Co.
Rogers, Minn. 55374
Vase toggles, harps, risers, swivels, socket covers, chimneys, brass tubing, figurine bends, locknuts, bushings, nozzles, finials, pipe, nipples, sockets, plugs, cord, swag chain

Span-O-Chain, Ltd.
7412 Bergenline Ave.
North Bergen, N.J. 07047
Chain

Topco, Inc.
107 Trumbull St.
Elizabeth, N.J. 07206
Balls 1 1/2" to 10", cans 2 1/6" to 5 3/4" diameter, 3'-19 1/4" high

Tubco Lamp Parts, Ltd.
1773 Bayly St.
Pickering, Ontario L1W 247
Canada
Pipes, arms, scrolls, lamp poles, hardware, sockets, lamp cords, chain

Ward Engineering, Inc.
13444 Wyandotte St.
North Hollywood, Ca. 91605
Swivels

Wescal Industries
18033 S. Santa Fe Ave.
Compton, Ca. 90221
Chains, harps

GENERAL SUPPLIERS: LAMPMAKER'S SUPPLIES; RETAIL (organized by state)

Alabama

Bryant Electric
2852 S. 18th St.
Homewood 35209

House of Lamps
1112 Telegraph Rd.
Prichard 36610

Mayer Electric
3500 5th Ave. S.
Birmingham 35222

SUPPLY SOURCES

Alaska
Acme Electric Co.
127 Minnie St.
Fairbanks 99701

Alaska Light and Electric
1515 Tudor Rd.
Anchorage 99502

Kennedy Electric Inc.
204 E. Fireweed
Anchorage 99503

Arizona
Phoenix Lamps
2225 E. Indian School Rd.
Phoenix 85016

Regal Ltg. Fixture Co.
5330 N. 12th St.
Phoenix 85014

Show Low Home Improvement
26 Deuce of Clubs
Show Low 85901

Tri-City Lighting Co.
2235 W. Broadway
Mesa 85201

Arkansas
Allen Lamp and Shade
2412 Broadway
Little Rock 72206

Fort Smith Ceramic Supply
7318 Rogers Ave.
Fort Smith 72901

The Lamp House
5608 R St.
Little Rock 72207

Smith Lamp Co.
US Highway 67 North
Corning 72422

California
Anthony's Lighting
9462 Magnolia Ave.
Riverside 92503

Babton's Lighting Center
1422 Santa Monica
Santa Monica 90404

Dooley's Hardware
5075 Long Beach Blvd.
Long Beach 90805

General Sales Corp.
745 20th St.
San Francisco 94121

Hali Spechts
8740 Wilshire Blvd.
Beverly Hills 90211

John J. Scott
1079 Shary Circle
Concord 94520

Michel's Electric
2285 Westwood Blvd.
Los Angeles 90064

Paul Ferante
8464 Melrose Ave.
Los Angeles 90069

Colorado
Artistic Lamp and Shade
4665 E. Colfax Ave.
Denver 80220

Dixson, Inc.
287 27th Rd.
Grand Junction 81501

Home Lighting
624 N. Tejon
Colorado Springs 80902

Korte
1129 Pearl St.
Boulder 80302

Connecticut
Indecor Inc.
9 Depot St.
Milford 06460

Lamp Fair Inc.
500 Boston Post Rd.
Orange 06477

Special Services
121 Hamilton Ave.
Stamford 06902

Tower Oschan
15 Dewey St.
Bridgeport 06605

Delaware
Betty Jo Cer Den
3722 Old Capital
Marshallton 19808

Columbia Electric Supply
Ocean at Cannon St.
Fenwick Island 19944

Dover Electric Supply Co.
1631 S. DuPont Highway
Dover 19901

Manor Hobby Center
Rt. 41-13 Penmart
New Castle 19720

Washington, District of Columbia
Columbia Lighting Fixture
635 K St., NW 20002

Eagle Electric Supply
940 New York Ave., NW 20001

House of Decoupage
4701 Sangamore Rd. 20016

Reed Electric
1667 Wisconsin Ave. NW 20007

Florida
Lamp Shade Fair
1336 N. Mills Ave.
Orlando 32803

Stewart Lighting
4751 San Juan Ave.
Jacksonville 32210

Uni Craft Corp.
4523 30th St, W.
Bradenton 33505

Georgia
Baker and Jarrell Electric
229 Price
Savannah 31401

Cliff Carter
1611 13th Ave.
Columbus 31901

Meriam Ezelle
2068 Walton Way
Augusta 30904

Putzel's Lighting
1343 Georgia Ave.
Macon 31201

SUPPLY SOURCES

Hawaii
Lindy Sales Ltd.
125 Mokauea St.
Honolulu 96819

Idaho
Gorden's Electric
805 N. Main
Moscow 83843

Key Building and Lighting Center
1036 Blue Lakes Blvd.
Twin Falls 83301

The Light House
830 Vista
Boise 83705

Wakefields
410 3rd St. E.
Ketchum 83340

Illinois
Courtesy Rand Rd.
700 E. Rand Rd.
Mt. Prospect 60056

Gearon Co.
3225 W. 26th St.
Chicago 60623

McAteers House of Lights
10808 Lincoln Terr.
Fairview Heights 62232

Springfield Electric
718 North Ninth St.
Springfield 61108

Town and Country Ceramic Supply
US Rt. 83 and Center
Grayslake 60030

Indiana
Keystone Lighting and Supply
6055 E. 82nd St.
Indianapolis 46250

Perry's Violin and Lamp
316-318 SE 8th
Evansville 47713

Sugar Creek Art Products
RR 8 Lafayette Rd.
Crawfordsville 47933

Thomas Electric Co.
1260 Jackson
Columbus 47201

Iowa
Antique Service Center
813 W. 3rd St.
Sumner 50674

The Lamp Shop
3215 Forest Ave.
Des Moines 50311

Lighting Associates, Inc.
2724 State St.
Bettendorf 52722

Stalker Electric
616 Second Ave.
Cedar Rapids 52401

Kansas
Evans Ceramic Supply
1518 S. Washington
Wichita 67211

Phillip Lakin
2040 Fort Riley Blvd.
Manhattan 66502

Rensenhouse Electric
9200 Marshall Dr.
Lenexa 66215

Rockwell Electric Co.
1111 Main St.
Goodland 67735

Kentucky
Daubert Electric Co.
1126 Bardstown Rd.
Louisville 40204

Hoffman Lighting Co.
7324 La Grange Rd.
Louisville 40222

The Lamplighter
2251 29th St.
Ashland 41101

Louisiana
C and C Electric Co.
1302 Louisiana Ave.
Shreveport 71101

The Lamp Post
2415 Government St.
Baton Rouge 70806

Meynier's Appliance Service
7911 Maple St.
New Orleans 70118

Nagem Electric Co.
933 Ryan St.
Lake Charles 70601

Maine
Elm Haven Farm Antiques
70 Portland Rd.
Kennebunk 04043

Aaron Hurewitz
Rt. 1
Stockton Springs 04981

Lewiston Supply Co.
71 Lisbon St.
Lewiston 04240

Smith's Ceramics
268 Main St.
Bangor 04401

Maryland
Gem Electric Supply Co.
11939 Tech Rd.
Silver Springs 20904

Lamps Unlimited
11610 Rockville Pike
Rockville 20852

Valley Lighting Co.
1010 York Rd.
Towson 21204

Wot-Not Shoppe
311 Crain SE Highway
Glen Burnie 21061

Massachusetts
The Art Shade Co.
165 Chestnut St.
Needham 02192

Joseph Leavitt Co.
6 St. Mark St.
Auburn 01501

Old Country Store
Elm St.
W. Mansfield 02083

SUPPLY SOURCES

The Whyte House Inc.
630 Worcester Rd.
Framingham 01701

Michigan
Brose Elec. Constr.
25825 Plymouth Rd.
Detroit 48239

Hudson's
14225 W. Warren Ave.
Dearborn 48126

Litscher Elec. Co.
910 Scribner NW
Grand Rapids 49504

Old Wayne Workshop
3455 Winifred
Wayne 48184

Minnesota
Brinkman Service Co.
286 S. Snelling Ave.
St. Paul 55116

Ceramic Arts and Supply
4634 Humboldt Ave. N.
Minneapolis 55412

H and H Electric
223 S. Broadway
Rochester 55901

Warner Hardware Co.
P.O. Box 17017
Minneapolis

Mississippi
American Elec. Mfg.
Stateline Rd. and Highway 55
Southaven 38671

Chain Lighting and Appliance Co.
1308 W. Pine St.
Hattiesburg 39104

Lamps Galore
203 S. State
Jackson 39201

Specialties, Inc.
98 W. Caroma St.
Olive Branch 38654

Missouri
Central Hardware
111 Boulder Dr.
Bridgeton 63042

Electrical Equipment Co.
404 South Ave.
Springfield 65806

Jaffe Lighting
823 N. 6th St.
St. Louis 63101

Townley Metal and Shade
P.O. Box 79
Kansas City 64141

Montana
CP Electric, Inc.
129 S. Main
Livingston 59047

Palmquist Electric Co.
420 N. Main St.
Helena 59601

Skyline Distributing Co.
P.O. Box 2053
Great Falls 59401

Yellowstone Electric Co.
520 N. 32nd St.
Billings 59101

Nebraska
D and M Ceramic
1620 W. Military
Fremont 68025

Jim's Gifts
419 Laramie
Alliance 69301

Henry W. Miller Co.
2501 St. Mary's Ave.
Omaha 68105

White Electrical Supply Co.
427 S. 10th St.
Lincoln 68508

Nevada
Bonanza Home Improvement
1715 Carson Rd.
Carson City 89701

House of Lamps, Inc.
3060 E. Fremont St.
Las Vegas 89105

Vaughn Materials
P.O. Box 679
Reno 89104

New Hampshire
Ormsbee's Lamps
Rear 54 S. Main St.
Concord 03301

Norman Perry Inc.
Box 90 Railroad Sq.
Plymouth 03264

Towne House
30 Ellison St.
Jaffrey 03452

Yield House Gift Shop
Junction Rts. 16 and 302
North Conway 03860

New Jersey
Commercial Lighting Products Co.
1500 Suckle Highway
Pensauken 08110

Cross Keys Metalcraft
251 Blackhorse Pike
Turnersville 08012

Flynn's Fromm Electric
808 Haddonfield Ave.
Cherry Hill 08034

Lampcrafters
97 Somerset St.
North Plainfield 07060

Nelson Lebo Co., Inc.
225 Monmouth St.
Trenton 08609

Surrey Electric
2432 Rte. 22
Union 07083

Williams Lamps
765 Central Ave.
Westfield 07090

New Mexico
Furman Galleries
110 E. Hope
Farmington 87401

House of Light
P.O. Box 15092
Rio Rancho 87114

The Lamp Shop
2320 E. Central Ave.
Albuquerque 87106

Tinnie Mercantile Co.
Box 100
Roswell 88201

New York
Buffalo Incan Light
201 E. Genesee St.
Buffalo 14203

Delta Supply
1268 Second Ave.
New York 10021

Limited Editions
253 E. 72nd St.
New York 10021

Lincoln Lite
761 10th Ave.
New York 10019

Maynard's Electric Shop
428 State St.
Rochester 14608

Schenectady Hardware and Electric
818 Albany St.
Schenectady 12307

North Carolina
Adams Wood Turning
216 Woodbine St.
High Point 27260

Summerour Lamps
900 E. Independence
Charlotte 28204

Thompson Lynch Co.
20 W. Hargett St.
Raleigh 27601

Wildwood Lamps
519 N. Church
Rocky Mount 27801

North Dakota
Electric Construction Co.
16 S. Fourth St.
Grand Forks 58201

Northwest Ceramic Supply
221 Broadway
Fargo 58102

Thunderbird Stores
623 N. Broadway
Minot 58701

Ohio
The Bostwick Braun
Summit and Monroe
Toledo 43602

Frankelite Co.
1425 Rockwell Ave.
Cleveland 44111

Industrial Electric Co.
5367 Northfield Rd.
Bedford Heights 44146

Mahoning Plumbing
946 Shehy St.
Youngstown 44506

Oklahoma
Albert's Electric
225 1/2 S. 15th St.
Fredrick 73542

American Electric Contr.
1105 N. Billen
Oklahoma City 73107

Davis Electric Center
4241 SW Blvd.
Tulsa 74107

Oregon
Brighter Homes Electric
1968 W. 6th Ave.
Eugene 97402

The Lamplighter
2937 E. Burnside
Portland 97214

Naomi's Cottage Shop
15942 Boones Ferry
Lake Oswego 97034

Pennsylvania
Austrian Lamp Co.
140 N. Second St.
Philadelphia 19106

Billows Electric Supply
7225 Frankford Ave.
Philadelphia 19126

Koch's Ceramics
624 Grove Ave.
Johnstown 15902

Lighting Pittsburgh
2855 W. Liberty Ave.
Pittsburgh 15222

Rhode Island
Breeze Hill Lamp Shop
1447 Wampandag Trail
E. Providence 02915

Cranston Elec. Supply
800 Oaklawn Ave.
Cranston 02920

The Lighting Center
207 Newport Ave.
Pawtucket 02861

Tops Electric Supply
124 Point St.
Providence 02903

South Carolina
Electric Wholesalers
303 S. Main St.
Sumter 29150

Levinson Electric
731 Meeting St.
Charleston 29403

Shops of Dainice
221 N. Main St.
Greenville 29601

Alvin D. Wall
1356 W. Wade Hampton
Greer 29651

South Dakota
Harry's Electric
1102 Jackson Blvd.
Rapid City 57701

Herter's Inc.
R.R. 2
Mitchell 57301

Van Dyke Supply Co.
Woonsocket 57385

Tennessee
Colonial Lamp and Supply
E. Arrow Dr.
McMinnville 37110

SUPPLY SOURCES

The Lamplighter
2107 Union
Memphis 38108

Mountain View Ceramic Center
1800 Dayton Blvd.
Chattanooga 37405

Williams Salvage Co.
127 3rd Ave. S.
Nashville 38103

Texas
A.B. Co.
209-219 S. Oregon
El Paso 79901

Bynum and Tanner Electric
2125 Franklin Ave.
Waco 76701

Enchanted Lighting
3121 Knox
Dallas 75205

Houston Arts and Crafts
2048 Marshall
Houston 77006

Utah
Century Light Center
259 31st St.
Ogden 84401

Duncan Electric Inc.
290 S. Main
Bountiful 84010

Streveil Paterson
1401 S. 6th St.
Salt Lake City 84102

Vermont
1836 Country Store
W. Main St.
Wilmington 05363

The Lamp and Shade
66 Pearl St.
Essex Junction 05452

Northern Electric Co.
83 Pearl St.
Burlington 05401

Virginia
Atlantic Electrical Supply
2117 Westwood Ave.
Richmond 23230

Dixie Pottery Warehouse
Rt. 2, US 11
Abingdon 24210

House of Lamps
108 S. Foushee St.
Richmond 23220

Lamplighter
1207 King St.
Alexandria 22314

Washington
Dor-Mick Ceramics
33350 Pacific Highway
Federal Way 98002

Hansen Lamp and Shade
6510 Phinney Ave.
Seattle 98103

Pacific Electric Co.
646 Fifth St.
Bremerton 98310

West Virginia
Cesco Lighting
1118 Main St.
Wheeling 26003

Electric Equipment Co.
473 High St.
Morgantown 26505

Erskine Glass and Mfg.
22nd and Lamplighter
Wellsburg 26070

Fenton Art Glass Co.
Caroline and 4 1/2 St.
Williamstown 26187

Smootz Lamp and Brass
200 N. Church St.
Charlestown 25414

Wisconsin
Century Hardware Co.
4711 Woolworth Ave.
Milwaukee 53218

Marcella Ceramic Studio
1150 Inman Parkway
Beloit 53111

Old Toll Road
16460 W. Bluemound
Brookfield 53005

Rolene Ceramic Studio
1593 Western Ave.
Green Bay 54303

Wyoming
Ace Electric Inc.
152 Grinnel
Sheridan 82801

Midwest Supply Co.
Box 2193
Casper 82601

LEATHER SUPPLIES

A.C. Products
422 Hudson St.
New York, N.Y. 10001
Leather

Berman Leather Co.
147 South St.
Boston, Mass. 02111
Leather

Tandy Leather Company Stores
Throughout the U.S.
Complete line of leather and supplies

World Art and Gift
606 E. State
Westport, Conn. 06880
Automatic Stitch Awl

MISCELLANEOUS

Bradley Enterprises
Main St.
Bradley Beach, N.J. 07720
Cork

Continental Felt Co.
22 W. 15th St.
New York, N.Y. 10011
Felt and felt parts

Dodge Cork Company
Lancaster, Pa. 17604
Cork

SUPPLY SOURCES

Kulicke Frames
43 E. 10th St.
New York, N.Y. 10003
Plastic and metal frame systems

Large department stores where home furnishings are sold
Baskets

Mechanical Felt and Textiles Co., Inc.
44 Fadem Rd.
Springfield, N.J. 07081
Felt, foam, foil laminates, buttons

Structural Industries, Inc.
96 New South Rd.
Hicksville, N.Y. 11801
Metal frame systems

The Wickery, Inc.
342 Third Ave.
New York, N.Y. 10010
Baskets

PLASTIC SUPPLIES

Abbion Cal, Inc.
123-21 AB Gray Ave.
Santa Barbara, Ca. 93101
Complete line of plastics and equipment

Ain Plastics
65 Fourth Ave.
New York, N.Y. 10003
Wide range of sheets, resins, and plastic findings, adhesives

Cadillac Plastics and Chemical Co.
15841 Second Ave.
P.O. Box 810
Detroit, Mich. 48232
Acrylic sheets, supplies, adhesives. Has branches throughout North America

Industrial Plastics
324 Canal St.
New York, N.Y. 10013
Wide range of sheets, resins, plastic findings, adhesives

The Plastics Factory
119 Ave. D
New York, N.Y. 10019
Wide range of plastics and supplies

Poly-Dec Co., Inc.
P.O. Box 541
Bayonne, N.J. 07002
Poly-Mosaic Tiles

Saks Arts and Crafts
207 N. Milwaukee Ave.
Milwaukee, Wisc. 53202
Assorted plastics, related supplies, and Poly-Mosaic Tiles

Alex Tiranti, Ltd.
70 High St.
Theale, Berkshire, England
Complete line of plastics and equipment

SHADE MAKERS
(For many more sources look in the yellow pages of your telephone book)

Jack Cushen's Glass Studio
260 West 29th St.
New York, N.Y. 10003
Leaded glass shades; gives classes in all glass arts

Lampcrafters
97 Somerset St.
North Plainfield, N.J. 07060
Custom-made lampshades—designing, restorations, repairs

Lasolite Corp.
718 Broadway
New York, N.Y. 10003
Leaded glass shades

Lite and Shade, Inc.
1193 Lexington Ave.
New York, N.Y. 10028
Designed and custom made shades

Oriental Lampshade Co.
810 Lexington Ave.
New York, N.Y. 10021
Ready-made and custom shades

Roseart Lampshades, Inc.
225 E. 134th St.
Bronx, N.Y. 10451
Hard and soft covered shades

Woodstock Workshop
214 Sullivan St.
New York, N.Y. 10012
Lampshades in fabrics, prints, and embroideries

SHADE-MAKING SUPPLIES

Artistic Bias Products Co., Inc.
31-33 West 21st St.
New York, N.Y. 10010
Trimmings, tapes, wraparound materials, braids, velvets, ruchings

Bienfang Paper Co., Inc.
P.O. Box 408
Metuchen, N.J. 07840
Lampshade parchment and general selection of papers

David Davis
530 La Guardia Place
New York, N.Y. 10012
Handmade and exotic papers

Dennison Manufacturing Co.
Framingham, Mass. 01701
Scorasculpture® for paper sculpture shades

Great Lakes Paper Co.
308 W. Erie St.
Chicago, Ill. 60610
Laminates, pleats, finding tapes, fabrics, parchments, foils, plastics, grass cloth, covers

Inter Design Craft, Inc.
P.O. Box 22128
23500 Mercantile Rd.
Cleveland, Ohio 44122
Wood slat hanging lamps in kit form—to be assembled and glued

Kirks Lane Lamp Parts Co.
1445 Ford Rd.
Box 519
Cornwells Heights, Pa. 19020
Student shades, chimneys, decorated glass

John Lewis and Co., Ltd.
Oxford St.
London W 1, England
Frames, firm coverings, soft coverings

SUPPLY SOURCES

Luger Manufacturing Corp.
500 Mill St.
Dunmore, Pa. 18512
All shapes of frames for lampshades, diffusers, chimneys, clip sets, cylinders

Paperchase Products, Ltd.
216 Tottenham Court Rd.
London W 1, England
Parchment, handmade and exotic papers

Primelite Manufacturing Corp.
407 S. Main St.
Freeport, N.Y. 11520
Ball globes of glass and plastic, with and without necks, 5-inch to 24-inch diameters

Riekes Crisa Corp.
1818 Leavenworth
Omaha, Neb. 68102
Glass shades, hurricanes, chimneys, etc.

Toolex, Inc.
State Highway 31
Ringoes, N.J. 08551
Complete range of rings, all shapes of frames, will custom-make frames and rings

STAINED GLASS SUPPLIES

S. A. Bendheim
122 Hudson St.
New York, N.Y. 10013
Stained glass importer, and complete line of equipment and supplies

Bienenfeld Industries, Inc.
1541 Covert St.
Brooklyn, N.Y. 11227
Stained glass and supplies

Blenko Glass Co.
Milton, W. Va. 25541
Stained glass

Claritude, Ltd.
19 Dunraven St.
Park Lane
London W 1, England
Glass

Glass Masters Guild
52 Carmine St.
New York, N.Y. 10014
Stained glass supplies, cardboard and asbestos molds for lampmaking

Glenside Glass
West Mt. Carmel Ave.
Glenside, Pa. 19038
Copper foil

James Hetley and Co., Ltd.
Beresford Ave.
Wembley, Middlesex, England
Glass

Stained Glass of Hanover
Whittlemore-Durgin Glass Co.
Box 2065
Hanover, Mass. 02339
Full line of stained glass supplies, molds for lampmaking

WOOD SUPPLIES

H. Behlen and Bros., Inc.
P.O. Box 698
Amsterdam, N.Y. 12010
Everything for wood finishing

Albert Constantin and Son, Inc.
2050 Eastchester Rd.
Bronx, N.Y. 10461
Veneers

Craftsman Wood Service
2727 South Mary St.
Chicago, Ill. 60608
Veneers, moldings, hardwoods, plywoods, supplies, tools

Designers Resource Group
Box 142
Seabrook, Tex. 77565
Massive cubes and cylinders

Elivilply Veneer Co., Ltd.
48A Eagle Wharf Rd., N. 1
London, England
Veneers

Freeman Products
250 Park Ave. S.
New York, N.Y. 10003
Turned wood elements, columns, tapers

Gaston Wood Finishes
3630 E. 10th St.
Bloomington, Ind. 47401
Everything for wood finishing

W. W. Howard Bros., Ltd.
Howard House Lanrick Rd.
Poplar
London E. 14 OJF, England
Veneers, hardwood, plywood

Little York Products
70 Hudson St.
Hoboken, N.J. 07030
Turned wood elements

Neill and Spanjer
Fairfield Ave. and Market St.
Kenilworth, N.J. 07033
Hardwoods, plywoods, moldings

Woodcraft Supply Corp.
313 Montvale Ave.
Woburn, Mass. 01801
Wood supplies, finishes, including Watco Danish Oil

Index

Italic figures indicate pictures.

Acrylic rod fixture, *174-75*
Acrylics. *See* Plastics
Adhesives
  ceramic tile adhesive, *109*
  Crazy-Glue®, *127*
  epoxy, 66, 106, *113*, 181-82, 184, 186-87, *191*, 229, 232
  ethylene dichloride, *127*
  methylene chloride, *121*, *127*, *133-34*, *139-40*, *158*, *173*, *191*
  mucilage and glycerine, *149*
  PS-30, *127*, *174-75*, *180-81*
  rubber cement, 87, *90*
  Silastic RTV (silicone), *145-46*, *191-92*
  solvent cement, *163*
  styrene cement (SC-1508), *147*, *195*
  Titebond®, 106, *191*
  Weldwood Plastic Resin Adhesive®, 106, *115*, *119*, *130-31*, *191*
  white, all-purpose glues (Velverette®, Sobo®, Elmer's®, etc.), 66, 72, 73, *78*, *83-84*, *93*, *94*, *97*, *101*, 106, *110-11*, *113*, *152*, *166*, *170*, *172*
Alcohol stains. *See* Finishing wood
Arms
  adjustable, 29, *29*
  flexible, 29, *29*
  See also Gooseneck lamp
Aulenti, Gae, *225*

Ballast (for fluorescent fixture), 58, *59*
Bases (for table lamps), 31, *31*, *33*, 62-63, *191*
  acrylic-clad, *229*
  ceramic, *16*
  marble, *153*
  plywood, *108-11*, *177*, *229*
  sewer pipe, *183*
  wood, *107*, *186*
Basket (shades and lamps), *102-3*, *168-69*, *183*, *186-89*, *193*
Beads (strung in shade), *103*
Bearman, Jane, *223*
Borden, Phillip, *223*
Bottle converter, *16*, *30*, *185*
Bottle lamp, *184-85*
Bulb socket. *See* Socket
Bulbs, 21, 58
  Christmas-type, *172*
  flame-shaped, *136*, *194*
  globe-shaped, *157*, *171*
  neon, *219*
  reflector-type, 22, 64, *137*, *148*, *153*
  smoky, *234-35*
  tubular, *114*, *148*, *229-30*
  two-in-one socket, 26
  various shapes, *16*
  See also Fluorescent lighting
Butterfly clip, *30*, *61*, *185*

Calder, Alexander, *217*
Campbell, John, *151-52*
Candelabra socket. *See* Socket
Canopy, 28, *33-35*, 56, *57*, *104*, *160*, *175*, *190*

Cantilever lamp, *16*, *168*
Carnuba wax. *See* Clear finishes
Castle, Wendell, *220*
Ceiling fixtures, *16*, 25-26, 34-36, 156-82
  assembly, 34
  wiring, 56-58
Check rings (seating rings), 26, *28*, 29-30, *30*, *32*, *111*
Chimney attachment. *See* Lampshades, attachments
Clear finishes. *See* Finishing wood
Coaxial cable, *180*
Color
  effects on mood, 22-23
  in stained glass, 196-98, 200
  of light, 13, 15, 22
  surface reflections of, 13-14
Conductance (of electricity), 42, 50
Connecting attachments, *191*
  See also Adhesives
Connecting wire, 26
  with screw terminals, *45*, *49*, *50*, 53, 55-56
  with soldering, 26, 48-49, *48*, *49*
  with solderless connectors (wire-nuts), *46*, 56, *57*, 58, 59
  with splicing, 46, *47*, 48
Construction, 25-42
  See also Electrical wiring; Lamp parts
Copper lampshade, *91*
Cork, *108-12*
Couplings, 28, 29-30, *30*
Crossbar, 34, *34-35*, *57*
Cushen, Jack, *v*, *197*, *199*, *201-7*

Decoupage, *84-85*
Die-cut Masonite panels. *See* Shoji lamp
Diffused lighting, *9*, 15, 23, 60, 87, *114*, *133*, *136-39*, *161*, *171*, *173*, *225*, *227*
Dimmers. *See* Switches
Distribution of lighting, 15, *17*
Dome lamp, *176-77*
Driftwood, *183*, *193*
Drilling holes
  in acrylic, *125*
  in glass or china, 184, *184*, 186
  in gourd, 190-91
  in wood, 107, 115, 119, 128-29, *131*, *147*
  in woven reed and fibers, 186-89
Drum lamp, *16*

Ecapuzzo, *225*
Electrical parts. *See* Plugs; Socket; Switches; Wire
Electrical safety, 26, 43-44
Electrical wiring
  connections, 45-49
  house circuitry, 42
  of bulb socket, 50, *51*
  of ceiling hookups, 56-58
  of clip-on plug, *50*
  of fluorescent fixtures, 58, *59*
  of in-line switches, 53-54, *53*, *54*
  of other switches, 54-56, *55*, *56*
  of series of sockets, *52*, *52*, *233*
  of standard plug, *49*, *50*

of wall hookups, 56-58
preparing the wire, 44, *45*
preplanning, 20, 26
safety, 26, 43-44

Faceted plastic sphere lamp, *157*
Felt, *132*
Figurine lamp, *16*, *107*
Findings (for lamp construction). *See* Lamp parts
Finials, 32, *32-33*, *103*, *190*
Finishes
  gold leaf, *151*
  Liquid Pearl®, *150*
  metallic finishes, *150*
Finishing wood, 122-23
  clear finishes, *120*, *123*, *131*, *233*
  sanding, *115*, *119*, 122, *148*
  staining, *107*, *111*, *113*, 122-23, *148*, *229*
Floor lamps, *16*, 25, *104-55*
  in mixed media, *152-55*
  in plastics, *123-52*
  in wood, *104-23*
Fluorescent lighting, *1*, 21, 58-59
  assembly, 58-59
  in light box, *221-23*
  lamps, *10*, *16*, *179-82*
  shade for, *64*
  starter switch, 40
  tube, 58-59
  See also Ballast
Fluted lampshade, *100-101*, *100*
Footcandles, 14
Found objects, 21, 26, *183-95*
  See also Baskets; Bottle lamp; Driftwood; Gourds; Pipe; Sconce
Fusible plastics. *See* Poly-Mosaics®

Gas lamp, 4, *6-7*, 9
Gauge, 42
Gimbal fitting. *See* Lampshades, attachments
Glare, 14, *230*
Glass
  in potichomania, *151-52*
  shade, *34*, *35*
  See also Bottle lamp; Globe lamps; Vase lamp
Globe lamps, *16*, *112*, *143-44*, *160-62*, *166-68*, *187*, *189*, *191-92*, *218*
Gluing wood. *See* Wood processes
Gooseneck lamp, *16*, *153*, *219*
Gourds, *183*
Grain matching (in shade material), 79
Greig, Dolores, *152*
Ground, 41

Hands lamp, *178-82*
Harps, 29, *33*, 62, *103*, *107*
  one-piece screw-on, *31-32*, *31*, *103*
  two-piece detachable, *30-31*, *31-32*, *103*, *190*
Hemp cord lamp, *156*, *166-68*
Hickey, 38, *38-39*, *188*, *193*

Illumination levels, 13-14
  and mood, 22-24

244

footcandles, 14
lumens, 14
See also Wattage
Incandescent lamps, 1, 7, 8, *17*, 21
See also Bulbs; and examples throughout
Insulation, 42-43
cardboard, 49
replaced with plastic tape, 49, *49*, *129*, *233*
replaced with Silastic RTV® (silicone sealant), 233
stripping of, *44-45*, 53

Joints (in woodworking), *108*
Junction box (ceiling box), 34-35, 56, 57, *161*, *187*

Krylon® (acrylic spray), 69, 84

Lacing, *74-75*
Lamp parts
availability, *x*
common to most lamps, *15*, 27-31
for ceiling lamps, 34-36
for table and floor lamps, 31-33
See also Electrical parts
Lampshades
anatomy of, *61*
and bulb size, 64, *78*
and use of harps, 31, *103*
applying soft coverings to frame of, 76, *77-78*
attachments, *61*, *62*
binding (wrapping) frame of, 69, 80, *82*, 87
burlap, *107*
clip-on, 31
cone, *61*, *62*, *63*, 67-68
copper, *91*
cylinder, *61*, *62*, *63*, 67-68, *116*, *129*
decoupaged, 84, *85*
design aspects, 20, 62-65
drum, *61*, *62*, 67-68, 84
empire, *61*, *62*
fluted, 100-101, *100*
from found objects, 65
function, 64
glass. See Globe lamps
hard material assembly of, 70-76
hexagonal, *62*, 68-69, *151*
hurricane glass, *152*
leather, 87, *87-91*
matchstick, *85*, *111*
materials, *61*, 63-64
hard, 21, *61*, *63*, 66-67, 70-76, 84
soft, 21, *61*, 65-67, 69, 76-79
oval, *61*, *62*, 70-73
overlay, 84
patternmaking, 67-69
pierced and cut, 86-87, *86*
pleated, 91-99, 101, *116*, *154*
preparing frame for, 60, *61*, 65, 69, 70
proportions, 18, 20, 62-63
rectangular, *62*, 68-69, *133*
scale, 64

shells, 102-3
slipcovered, 101
square, *62*, 68-69, 84
stretch jersey tube, *171*
styles, 60, *61*, 65
suitability of coverings for, 63-64
tailoring soft covered-type, 79, *79*
techniques of making, 65-83
trim, *73*, 76, *78*, 80-84, *81-82*
use of linings in, 69, 80, *80*, 87
wrapped, 101
See also Basket
Lapietra, Ugo, *228*
Latex, *176*
Lead caming, 198-99, 208-9, 214-15
Leather
beanbag lamp base, *9*
lampshade, 87, *87-91*
Light
designing with, 12-13
in art, *216*
physical attributes of, 1, 12
sources of, 1-11
Light boxes, 221-24
Lighting
appearance, 18
basic modes of, 15-18, *16*
development of, 1-12
economy, 21-22
for specific activities, 14
mood, 13-14, 22-24, *216*, *218*
practicability of maintenance of, 18-20
Linings. See Lampshades, use of linings in
Load (on a switch), 55-56

Matchstick shade. See Lampshades
Merrill, Virginia, *85-86*, *151*
Mirror lamps, 121-22, *148*, *155*, *173-74*, *179-82*, *219*, *229-30*
Morrow, Lewis, *151-52*

Newman, Jack, *178*
Newman, Lee S., *178*, *230*
Newman, Thelma R., *219*, *221*, *224*
Nipples (steel), 28, *28*, *30*, *33*, 34, *57*, *121*, *135*, *140*, *147*, *193*
Nuts, 26, *28*, 29, *30*, *30*, *33*, 34, 56, 57, *107*, *111*, *121*, *135*, 189-90, *192-93*
Nylon, strung on acrylic frame, *156*, *158-60*

Oil stains. See Finishing wood

Pan lamp, *16*, *173*
Paper
in shade making, 66, 91-101
in Shoji lamp, *113-14*
Japanese lanterns, *140*, *172*
Parchment shades, *61*, 70, *85*, 86-87, *86*, *112*, *137*
Paris-Craft®, *178*
Patternmaking (of lampshades), 67-69
Pigmented wiping stains. See Finishing wood
Pipe
aluminum, *231*

brass, 28, *28*, *30*, *32*, *32*
chrome, *137*, *154*
threaded steel, 27, *27-28*, 29-30, *30*, *51*, *57*, *107-8*, *111*, *121*, *128-29*, *190*, *192*
Pipe lamp, *231-36*
Plaster of Paris, *178-79*
Plastics
acrylic, 123-24
and neon lights, *219*
and wood scraps lamp, *125-33*
bending, 141-43
cementing, *121*, *127*, *128*, *133-34*
cylinder, *145*, *158*
dome lamp, *176-77*
drilling, *125*, *228*
etched, *223*
extruded rods, *140*, *174-75*
finishing, *125-27*, *134*
gesso, *152*
heating and forming, 141, *141-43*, *160*, *162*
hurricane lamp, *133-36*
marking and sawing, *121*, 124-25, *133*
modeling paste, *146*
paints, 84, *107*, *176-77*
piping of light, *10*, 222-28
polishing, *125*, *126*, *127*, *134*
rectangular tube lamp, *138-39*
sanding, *125*, *126*
scraping, *125*
tapping, *125*
techniques, 123-52
and wattage, 21
faceted sphere lamp, *157*
fusible plastics. See Poly-Mosaics®
in shades, 66, 70, 74, *85*, *91*, *118*, *121-22*
Mylar® tape, *135*, *139*
polyester resin and fiberglass lamp sculpture, *220*
polyethylene spheres (in globe lamps), *143*
styrene cup lamp, *195*
vinyl, *9*, *156*, *162-65*, *219*
Plasti-Tak®, 87, 199-200
Pleated lampshades, 91-99, 101, *116*, *154*, *190*
attaching, *93*
forming, *92*
variations, 94, *94-99*
Plugs, 41
clip-on, *50*
standard, *49-50*
Pole lamp, *16*, *18*, *104*, *119-22*
switch for, 56
Poly-Mosaics®
cutting, *211*
cylinder lamp, *145-48*
fused lamps, 208-15
fusing, 208, *210-11*
Poly-Web®, 66, 73
Portable lamps. See Floor lamps; Table lamps
Potichomania, *149-52*

Reflectance, 13, 60, *137*

Reflector bulbs. *See* Bulbs
Risers, *32, 103*

Sconce lamp, *16, 193-94*
Scrap lamp (acrylic and wood), *125-33, 133*
Sculptural lamps, *216-36*
    light boxes for, *221-27*
    mirror lamp, *229-30*
    pipe lamp with smoky bulbs, *16, 231-36*
Seating rings. *See* Check rings
Series wiring. *See* Electrical wiring
Shades. *See* Lampshades
Shellac. *See* Clear finishes
Shoji-style lamp, *113-14*
Slipcover lampshade, *101*
Socket (bulb socket), *27-28, 31, 32-33, 36-39, 107, 111, 114, 121, 135, 138, 140, 153, 171-72, 177, 185, 189-190, 192-93, 229, 234*
    base sizes, *36, 112*
    clip-on type, *52*
    finishes, *36*
    twin light adapter (clusters), *36, 39, 188*
    wiring, *50-51*
Soldering, *26, 48-49, 129, 183, 204-6, 209-11, 213*
    flux, *48, 204, 209-10*
    iron, *48, 48, 204*
    technique, *48-49, 48, 204-6, 209-10, 211, 213*
Spider, *189*
Splicing, *26, 46-48, 46, 129*
    at right angles, *47*
    pigtail, *46, 56*
    Western Union, *47*
Stained glass lamps, *8, 196-215*
    cutting glass for, *200-202*
    patternmaking (template), *198-200*
    soldering, *204-6, 209-10*
    Tiffany-style, *8, 198*
    types of glass, *200*
    wrapping glass with copper tape/foil, *203, 209-10*
Stephens, Curtis, *v, 162-64*
Stretch jersey tube lamp, *171*
Strip heating element, *141, 141-42*
    See also Acrylic, bending
Strip lamp, *16, 137*
Swag lamp, *16*
    canopy for, *34-35*

Switches
    built into sockets, *36, 37-38*
    dimmers, *36, 54, 155, 194*
    fluorescent starter, *40*
    in light box, *212*
    in-line, *39, 40-41, 53, 53-54, 139*
    pull chain, *40, 40, 55*
    push button, *40, 40, 55, 180*
    rotary, *40, 40, 55, 112, 122, 128*
    three-way rotary, *40, 40-41, 56*
    toggle, *40, 40, 55*
    See also Electrical wiring
Swivel, *129*

Table lamps, *25, 104-55*
    assembly, *31-32, 32-33*
    desk lamps, *10, 142, 153, 182*
    in mixed media, *152-55*
    in plastics, *123-52*
    in wood, *104-23*
    parts, *31-33*
    proportion to shade, *62-63*
Texturing wood, *106-7*
Threaded steel pipe. *See* Pipe, threaded steel
Tiffany Lamps. *See* Stained glass lamps
Tools
    bone folder, *92, 110*
    burnisher, *203, 212*
    chisel and mallet, *87, 128, 147, 233*
    clamps, *114, 116, 119, 128, 131, 134*
    crosscut saw, *115*
    drill, *107, 115, 119, 128-29, 147, 184, 187*
    electric saws, *121, 123, 124, 133, 147, 180, 182, 211*
    file, *27, 69, 231*
    for texturing wood, *106-7*
    for woodworking, *105*
    glass cutter, *200-202, 202*
    grozing pliers, *201, 202*
    hacksaw, *27, 28, 231*
    hammer, *89, 91, 108, 190*
    hole punch, *74*
    knife, *44, 81, 86, 88, 92, 97, 109, 110, 113, 132, 162*
    lacing needle, *75*
    leather stitching needle (lockstitch awl), *87, 89*
    needle, *76*
    plastic syringe, *121, 128*

pliers, *45, 49*
propane torch, *213*
razor blade, *44, 111*
router, *129, 135*
running pliers, *201*
sandpaper, *47*
scissors, *71, 79, 82, 88, 90*
screwdriver, *50, 51, 57, 59, 119, 232*
sewing machine, *101*
spacing wheel, *88*
staple gun, *99, 129*
strip heating element, *141-42, 160*
Surform® tools, *106*
tile cutters, *146*
vise, *27, 231*
wedge, *206*
wire stripper, *44*
Trimming. *See* Lampshades
Tubing. *See* Pipe
Turned wood lamps, *115-18*

Underwriter's knot, *49, 50*
Uno-bridge. *See* Lampshades, attachments

Varnishes. *See* Clear finishes
Vase lamp, *186*
    See also Bottle lamps; Potichomania
Vase toggle, *33*
Vellum, *61, 66, 85, 86, 100*
Veneering, *106, 108, 110-12*
Vinyl scored lamp, *162-65*
    See also Plastics
Visual comfort, *13-14*

Wall fixtures, *16, 25-26, 156-82*
Washers, *28, 29-30, 33, 107, 111, 121, 189, 234*
Water stains. *See* Finishing wood
Wattage, *14, 21-22, 64, 177, 185*
Wire, *41-42*
    See also Electrical wiring
Wire nuts (solderless connectors), *46, 56, 57, 58, 59*
    See also Connecting wire
Wiring. *See* Electrical wiring
Wood
    processes, *106-23*
    tools, *105*
Wood strips lamp, *169-70*